MW01618290

Chemical Cleaning Disinfection and Decontamination of Water Wells

John H. Schnieders Ph.D.
Certified Professional Chemist

Illustrations

Michael J. Schnieders
Karen J. Couch

Published by Johnson Screens Inc., St. Paul, MN 55112

Library of Congress Catalog Card Number 2003104706

ISBN 0-9726750-0-0

First Edition, First Printing

Printing Enterprises, Incorporated printed this volume on sixty-pound Starbrite stock. Typography was done by Infinity Direct, Incorporated; the text is set in 12 point Times Roman type face.

The information and recommendations contained in this book have been compiled from sources believed to be reliable and to represent the best opinion on the subject as of 2003. However, no warranty, guarantee, or representation, express or implied, is made by Johnson Screens as to the correctness or sufficiency of this information or to the results to be obtained from the use thereof. It cannot be assumed that all necessary warnings, safety suggestions, and precautionary measures are contained in this book, or that any additional information or measures may not be required or desirable because of particular conditions or circumstances, or because of any applicable U.S.A. federal, state, or local law, or any applicable foreign law or any insurance requirements or codes. The warnings, safety suggestions, and precautionary measures contained herein do not supplement or modify any U.S.A. federal, state, or local law, or any applicable foreign law, or any insurance requirements or codes.

Preface

This book was written for water well professionals and public health officials who find themselves having to decontaminate, disinfect, or clean a water well. The well may have a contaminant problem due to the presence of indicator organisms such as coliforms, sulfate-reducing bacteria, or a known pathogen. The contamination may have occurred from natural catastrophes such as flooding, earthquakes, water shed occurrences, or it may have resulted from accidental pollution or more willful acts. The problem may not be a contaminant but just the natural buildup of mineral and biological blockage resulting in corrosion, water quality change, or production loss. The primary objective of this book is to provide knowledge of how microorganisms inhabit, minerals accumulate, and how cleaning efforts should be directed for maximum effectiveness. This book may also be used as a general guide for the cleaning of well systems contaminated with any chemical or biological substance.

The book is divided into several sections. Chapter 1 contains recommendations for disinfection and Chapters 2 through 6 are a review of current methods of disinfection and the biology and mineralogy of wells. Chapters 7 and 8 cover geological considerations and laboratory testing. The chemistry and methodology of well cleaning including chemical selection and mechanical methods are contained in Chapters 9 through 11. Neutralization of cleaning and disinfection chemistry is covered in Chapter 12 and specifications and case histories are provided in Chapters 13 and 14.

The methods mentioned in the following chapters were gleaned from numerous field applications by many contractors and water well professionals from around the country and the world. Integrated into these applications is the general research work of my laboratory colleagues at Water Systems Engineering, Inc. The primary goal of our more recent work was the removal of coliforms and the chemical cleaning of the microbiology and mineralogy that leads to coliform retention in wells. The general methods of decontamination were borrowed from all the work which has been done on understanding and developing the disinfection process. Early on it was discovered that the requirements of a successful disinfection were the same requirements for successful removal of mineral and biological blockage in wells (contaminants themselves).

As the coliform organism was studied to perfect the disinfection process, certain hiding techniques or habitation profiles became apparent. These characteristics which make the elusive coliform so difficult to remove are found in most of the more probable organisms that may contaminate well systems, particularly bacterial pathogens.

Finally, as the disinfection and decontamination process for organisms evolved, we began to use various aspects of the procedures in the removal of all types of contaminants from wells. Therefore the methods and procedures given in this book should be useful in the "decontamination" and removal of all foreign chemistries or microbial contaminants which may become entrenched in our water well systems.

The Author

John H. Schnieders is a Certified Professional Chemist and the principal chemist at Water Systems Engineering, Inc., a consulting and research laboratory specializing in the microbiology and chemistry of water. He has worked in the water industry for over 35 years, principally in the areas of microbial and chemical induced corrosion, blockage, and water quality loss. He holds degrees in Microbiology and Chemistry and a doctorate in Environmental Engineering.

John began working in groundwater about 18 years ago and has directed his laboratory toward the study of the chemical and microbial processes that make up the daily life of a water well. His team has developed not only knowledge of how well blockage occurs but better methods for the chemical cleaning, disinfection, and the general chemical rehabilitation of the well environment. His goal and that of this book is to apply basic principals of microbiology and chemistry to the chemical cleaning and maintenance of water wells.

John can be reached at his laboratory:

Water Systems Engineering, Inc.
P.O. Box 700
Ottawa, Kansas, USA 66067
Internet Web Site: http://www.h2osystems.com

Acknowledgments

This book would not have happened if the many people of the groundwater industry had not shown me the support and encouragement that I have received over the last twenty years. I cannot imagine an industry as alive and vibrant and yet unilaterally helpful as ours. I do have a number of very special people to thank. First, my lovely wife for her many hours of proofing and reproofing, not to speak of the thousands of phrases and ideas she had to listen to long before they made it into the book. My son, Michael, and my office manager, Vicki Staley, invested heavily for me. Vicki put in many hours of typing, proofing and retyping of the manuscript. Mike, who has his masters in hydrogeology and is also a practitioner of the art of well rehabilitation, discussed endlessly the many facts and theories that went into the writing of the book and prepared many of the well illustrations, for which I am indebted.

Thanks also to Karin Ybarra, our lab manager and excellent microbiologist, Craig Adkinson, chemist, Kathy Wiseman, environmental microbiologist, Connie Rich, chemist associate, Jason Kish, molecular biologist, Pat Adkinson, chemist, Michelle McClay, Elvin Currant, and Joe Ream who make up the excellent staff at Water Systems Engineering, Inc. Again, without their support and/or technical assistance the book would not have been possible.

Thanks to the great people at the NGWA who have encouraged me over the years and the people at Johnson Screens who introduced me to groundwater many years ago. Their continued support through the years has always been an inspiration. I need also to mention some very important people who have mentored me through the years: Mike Mehmert of Johnson Screens and Roger Miller with Layne Christensen. Roger Miller has been a friend and fellow chemist all the way back to the days of water plant operation before we entered the groundwater environment. Mike has been my major professor in the areas of geology and well construction. We have had an ongoing friendship for 15 years during which he has introduced me to and explained many phenomena of geology and the well. Last, but not least, I thank my reviewers for their many hours of anguish and, I'm sure, hesitancy as they poured over the pages: Jim Baird, Fletcher Driscoll, Sam Gershon, Dave Kill, Mike Mehmert, and Roger Miller.

Disclaimer

This book contains information on the chemical cleaning disinfection, and decontamination of water wells. It provides suggestions both in formula format and as recommended products that our lab has researched. These suggestions are not a substitute for experience and education. While we are not able to guarantee the quality of the various products we have recommended, we have found them to be some of the best on the market as a result of our research and field use.

Cleaning and disinfection of water wells should be supervised by professionals. If you are a well owner and require a consultant to specify the required work or contractors to perform the work, this book should help you understand the process and help them do a better and safer job. Water well professionals should find the book helpful in explaining many of the natural phenomena of well operation and cleaning as well as a review of the many chemistries available to them. The book should also provide a firm foundation on which to build their own techniques and methods.

Table of Contents

List of Tables

List of Figures

CHAPTER 1 Current Methods

The terms disinfection, decontamination, and chemical rehabilitation are all terms that, when used in reference to water wells, denote a form of cleaning.

Disinfection usually refers to the removal or destruction of microorganisms that are deemed contaminants but are usually coliforms, an indicator of possible contamination.

Decontamination refers to the removal or destruction of either microorganisms or chemical contaminants of a well and is usually the term specifically reserved for more disaster related contamination or pollution of wells. This is especially true if known pathogenic microorganisms or highly toxic chemicals are involved. Contaminants, however, can be any products or microorganisms or unusual accumulation of those products or microorganisms that are interfering with the operation and/or water quality of the well. Thus, the buildup of typical environmental organisms that may inhabit a well and/or the accompanying mineral accumulation are contaminants.

Chemical rehabilitation is used to refer to the systematic "cleaning" of a well to remove the mineral and bacteriological blockage (contaminants) which has resulted in the loss of capacity and/or water quality. There are many levels of rehabilitation with some activity (or steps) in the overall process alone deemed rehabilitation. Super chlorination is one of those "steps" often referred to as rehabilitation. Little more than simple chlorination is usually practiced when well owners are faced with true contamination because of the lack of any uniformity or specification. The more severe the contamination is deemed, the higher the concentration of chlorine used. Thus, while we think of the chlorination procedure as a cleaning mechanism, we are really only trying to "kill" the contaminant and have not used a procedure that is designed to clean the well and remove any debris that may be resulting in the contamination or adding to the problem.

Simple Chlorination or Disinfection

If the chlorinating procedure does not involve some actual cleaning, then the end results are usually less than expected. An example is the routine disinfection of water wells which have tested positive for coliforms. This procedure is usually called simple chlorination. The positive coliform test evokes the pouring or placement of a hypochlorite solution into the well to "disinfect" the well. A minimum concentration is required for a set time period; however, usually a much higher level of chlorine is used. Most of the time this results in failure or subsequent testing shows positive for coliform bacteria. This book will show the need to incorporate a cleaning step in the disinfection process or that true decontamination of a well system must include a cleaning procedure.

Over the past few years more chlorination has been used because the need for disinfection has increased due to the larger group of groundwater source supplies now under regulations. More regulation of existing wells and improved testing procedures have resulted in more positive coliform tests. As more attention has been given to the rechlorination procedures required, more information has been collected on those steps that have been successful. This development or positive change has slowly resulted in more successful disinfection procedures.

Simple chlorination, which implies only the addition of chlorine in the form of sodium hypochlorite or calcium hypochlorite, or in the case of home wells, household bleach or HTH, has often failed. Failed not because the method has become less effective, but because the test is considerably more sensitive. In many cases frustrated contractors and owners have added more and more chlorine and even then these astronomical concentrations of hypochlorite failed to "clean" the well.

The cases for the larger wells have not fared any better. Many contractors who are unable to turn municipal wells back over to their owners have repeatedly chlorinated to levels in excess of 5,000 mg/l. Considerable money is spent by the contractor or well owner who has to get the well on line. State and local public health departments have tried to help but their hands are tied by the regulations that deal with the presence of coliform bacteria in source water.

Gradually the tide is turning. Contractors who have removed the pump and cleaned out the casing before adding the chlorine have seen improvements, and those who have worked the chlorine solution into the well gravel pack or formation using a jetting tool or surge block have had even more success. Use of the chemistry that improves pH control of the chlorine solution and better control of concentration and volume have elevated the methodology. Many groups have discovered that better development of wells during construction, or full redevelopment following pump work or other maintenance, improves the success rate of chlorination considerably.

Research on Biofilms and the Chlorination Process

Research done on biofilm organization and reported at the International Conference on Microbial Biofilms in 1996 illuminated the role of biofilms as a protective device for certain bacteria. Marsh and Bradshaw (1996) showed that certain fastidious microorganisms (such as coliforms) are often recovered from multi-species biofilms in environments which appear hostile to their growth. Their research showed that coaggregation occurred between facultative species and anaerobes and that this coaggregation may enable small, metabolically structured units to form within biofilms. These units could then enable fastidious species to survive and proliferate within biofilms under otherwise stressful conditions.

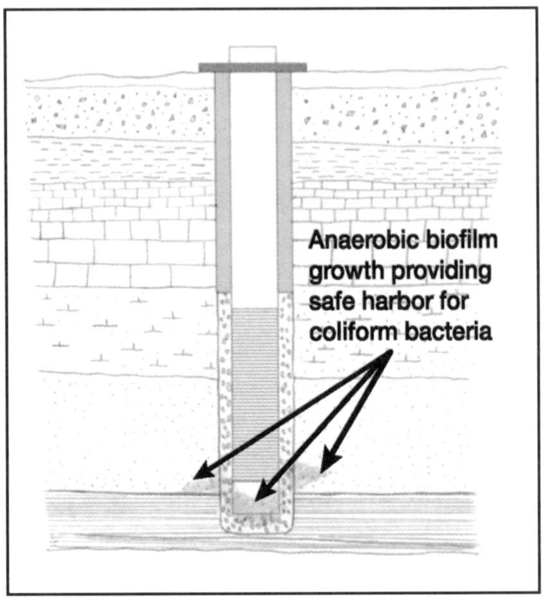

Figure 1.1. Gravel pack well showing anaerobic area Anaerobic area is created from low flow and debris accumulation.

Subsequent laboratory studies on biofilm formation in the well environment showed that following chlorination only deep, predominately anaerobic biofilms harbored coliform bacteria (Schnieders J., 2001). These findings were helpful in explaining the increased effectiveness of chlorinating procedures that are designed to reach deep biofilms usually located in the anaerobic zones of the well (Figures 1.1 and 1.2). Once the habitat or "hiding place" for coliforms was recognized, procedures could be devised to improve the disinfection process.

Work done in Michigan through the Department of Environmental Quality (DEQ) and our laboratory in Kansas resulted in the following recommendations for improved chlorination. (This work was also verified in many instances across the states in both large and small wells and with many different contractors. Many well professionals shared their findings to develop these recommendations.)

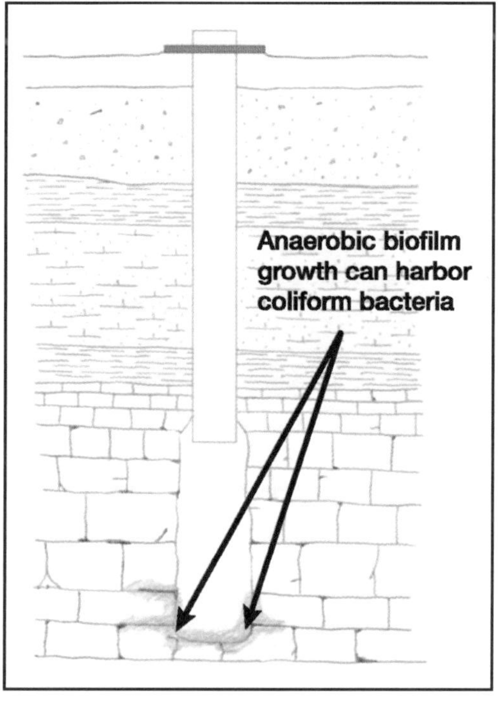

Figure 1.2. Open borehole well showing anaerobic area
Lower zones in fractured rock are ideal areas for anaerobic growth.

Recommendations for Water Well Disinfection

1) With new well construction or rehabilitation sufficient time should be spent on well development. With older wells you must remove mineral and biological accumulation prior to the addition of chlorine or any other cleaning chemistry.

2) For best results pump the well for an extended period of time before disinfection (24 to 48 hours for wells five inches or smaller and as much as two to four weeks for larger wells).

3) Prepare a chlorine solution (between 50 and 200 mg/l) above ground in sufficient volume to equal four times the standing well volume. The chlorine solution should be prepared by first adjusting the pH with an acid or

with a suitable acid product such as ChloraPal by Design Water Technologies or NuWell-310 or NuWell-410 by Johnson Screens, Inc. (see Chapter 9). Once the pH of the water reaches 4.5 to 5.0 add the sodium hypochlorite and mix.

WARNING: This must be done in a well-ventilated area out in the open as chlorine gas can be produced. Chlorine is heavier than air so vaults or low areas should also be avoided. Once the sodium hypochlorite is added, the pH will immediately rise and prevent chlorine release. (Chlorine is not released above a pH of 5.0.)

4) Place the prepared chlorine solution into the well, preferably with the use of a tremie line, adding it first at the bottom and releasing more solution as the line is raised. Surge or swab the solution into the gravel pack or immediate zones around the well. In wells with open borehole construction, jetting the solution may be preferable to the use of a surge block or swab. The application time is dependent on how deep the well is, but try to spend one to two minutes for every foot of well screen or open borehole below the water level. Some cleaning should be done above the screen to make sure that casing surfaces are disinfected. Let the solution stand over night.

5) The second day surge or jet for one-half the original time and evacuate the well starting just below the static water level moving downward in even increments toward the well bottom. The more vigorous the evacuation the better. Evacuate until no chlorine is detected but total volume discharged should equal at least 20 well volumes.

The changes initiated by the laboratory work, field surveys, and the successful implementations over a wide range of well sizes, environments, and contractors have brought about improvements in all types of well rehabilitation and cleaning projects. The information derived from studies of anaerobic sites deep in gravel packs or formations has illuminated the process of mineralization as well as the formation of heavy biological structures that block flow pathways in the vicinity of wells. This information has been used to develop special dispersants that have increased acid cleaning efficiency in well rehabilitation projects. Other dispersants have been studied that are more specific toward the bacterial exopolymers that make up 85% of the biomass. Since more than 80% of well plugging is believed by most well professionals to be caused by biofouling, more studies are needed to further understand the many chemical and biological reactions that take place in the well environment.

Chemical Rehabilitation

Besides chlorination, wells often undergo another form of cleaning referred to as chemical rehabilitation. This usually involves removing the pump, injecting a mixture of acid and a surfactant and/or dispersant into the well after which a wide variation of physical activity then takes place. Some well contractors only let the well set for a specific period and then pump the chemical from the well. Others, who may have left the pump in the well, will "bump" the pump in an attempt to mix the chemistry and force some of it into the zones around the well. Another method is to circulate the chemistry out of the well and return it in an attempt to push chemical out into the surrounding zones by the force of the chemical solution falling back into the well. In larger well systems more sophisticated activity is employed. Wire brushing of the casing and screen areas with evacuation of the debris takes place before the chemicals are added. Once the chemistry is ready to be placed even more ingenuity is used. Often the chemistry is tremied into the areas of the most accumulation as determined by videoing the casing and screen areas. While use of the video camera adds some degree of increased knowledge, it usually only shows what is on the surface of the well interior. No indication of gravel pack involvement is determined. In addition to treming into the affected region, packers are often used to isolate the cleaning activity and allow more force or energy to be directed at the cause of the blockage.

Methods Useful for Decontamination

Some techniques which are necessary for the successful removal or destruction of various pathogens and/or chemical toxins following natural or man-made disasters are the same methods used at various times in the cleaning, disinfection, and the rehabilitation of water wells. This book will attempt to illustrate those methods and the reasons why they are successful in removing coliforms, mineral and biological blockage, and contaminants from wells. Additionally, the book will show the methods of habitation most probable for many known pathogens and why these techniques can be successful for their removal. Finally, the book will give guidelines for well cleaning that can be successful for the neutralization or removal from the well of toxic material more chemical in nature. In this case, the book will show the procedures and make suggestions as to probable chemistry which can be substituted with other chemistry specific to the neutralization or dissolution of a particular toxic material.

CHAPTER 2 Contaminants in Well Systems

Within the scope of this book a contaminant is defined as any microorganism or chemical substance that is either foreign to the well or by the nature of its accumulated size is interfering with the operation or water quality of the well. Indicators of such contamination would also be included. Procedures outlined in the book can be helpful for the more routine contaminants such as coliforms or the sulfate-reducing bacteria, which cause taste and odor problems, as well as *E. coli,* which is a known human pathogenic or disease causing bacteria. They should also provide guidance for the cleaning of wells that have become contaminated as a result of natural disasters, accidents, intended or overt acts, or the results of natural occurrences. Such incidents often result in contamination from oils, agricultural, and industrial chemicals as well as fuel contamination. Several years ago a worker inadvertently pumped a couple of drums of spent motor oil into a well system. It is for just such an incident, or perhaps the contamination following a flood, that this book could provide a basis for an emergency cleanup program. It should also provide a foundation for the development of a program for the disinfection or decontamination of any toxic chemical or microbe that comes to inhabit or reside in a well system.

Microbial Contaminants

Microbial contaminants are usually thought of as microorganisms that are foreign to the aquifer water. Since so many microorganisms are already part of the natural population of aquifers, most generally organisms that are responsible for disease are the ones deemed contaminants. The coliform which has been designated a contaminant by regulatory default is actually an indicator organism used to identify the possibility of pathogenic microorganisms which are harbored in wastewater. While the number of possible pathogenic microorganisms is large, the necessity of developing a separate protocol for each bacterium is not necessarily required.

Potential microbial contaminants can be found among most groups of microorganisms; the viruses, the bacteria, the protozoans, and, we may find in the future, certain protein particles. Without producing a list of actual organisms, let's look at different groups of bacteria and how we expect them to exist in the well environment. Organisms can be divided into those that require oxygen to live, aerobes, or those that live without oxygen, anaerobes. There are large groups of bacteria however that are facultative; that is, they can live in either environment. Most protozoa require oxygen to live while viruses can live in either environment. Protozoans usually derive their energy by feeding off bacteria while viruses replicate by living inside the bacteria. To further complicate the bioenvironment many bacteria live in symbiotic relationship with other bacteria. This relationship often allows them to live in what otherwise would be considered a hostile environment.

In the study of well biology, a particular biological structure plays a commanding part in complicating the well environment. This structure is made up of bacteria and the extra cellular polymeric substance (EPS) they produce and is known as a biofilm. In researching how biofilms develop in the well environment, it became apparent that while biofilms harbor aerobic and anaerobic bacteria, certain types of biofilms dominate in the environment most suited to their oxygen requirement. We find, therefore, biofilms with predominately aerobic bacteria in areas of high oxygen potential. These areas, like the casing and screen zones, as well as gravel packs under the influence of aerated or oxygenated water, harbor the iron oxidizers and many of the slime formers such as the *Pseudomonas, Aeromonas, Acinetobacter, Flavobacter, Arthrobacter,* etc. It is in the areas of less or no oxygen, the lower zones or areas under clay lenses or heavily impacted gravel packs, and other anoxic formation zones that we find the predominately anaerobic populated biofilms. The rest of the bacteria present in the well system, but not included in the biofilm, are in the water (free-swimming) and are called planktonic bacteria.

Figure 2.1. Artistic rendering of biofilm
Artistic rendering of a series of micrographs of biofilm from studies at the Institute For Biofilm Research at Montana State University. Note the mushroom-like structures and the presence of several types of bacteria.

The removal of microorganisms or the decontamination of the well must take into consideration how the organisms reside in the environment. Those organisms that are free-swimming in the flowing water will wash from the well under normal operation or pumping. At the very least, low levels of chlorination will more than control these organisms. The organisms contained in the biofilms will be the most difficult to remove. Understanding this biofilm and the techniques and chemistries affecting it will provide the key for a successful decontamination or cleaning.

If we consider the free swimmer easiest to remove and realize that cleaning chemistry will add to their removal rate, our main concern must then be the entrenched organisms in the biofilm. Protozoans would feed in the biofilm, most probably in the aerobic areas; however cysts could accumulate in areas of less flow and therefore be found also in the anaerobic zones. Viruses, while in the main flow of water, will most probably be more concentrated in the areas of heavy biofilm activity either as the result of an accumulation due to the filtration effect of the biofilm structure or the presence of host organisms.

Most contamination incidents of a microbiological nature occur from the ingress of pathogenic bacteria and viruses usually of organisms found in the discharge of human or animal waste. The bacterial group would be dominated by facultative and anaerobic bacteria. Indeed, at least 80% of the list (Most Probable Pathogens that Can Contaminate Our Water, published by the American Academy of Microbiology in 2000) are organisms that fit the profile of those that we could expect to find in the anaerobic biofilms (Table 2.1).

Table 2.1. Most probable bacterial pathogens that can contaminate water

SPECIES	CLASSIFICATION
Acinetobacter calcoaceticus	Aerobic
Aeromonas hydrophila	Facultative
Aeromonas sobria	Facultative
Aeromonas caviae	Facultative
Bacteroides fragilis	Anaerobic
Bacteroides melaninogenicus	Anaerobic
Brucella spp.	Aerobic
Campylobacter fetus	Microaerophilic
Campylobacter jejuni	Microaerophilic
Chromobacterium violaceum	Facultative
Citrobacter spp.	Facultative
Clostridium botulinum	Anaerobic
Clostridium difficile	Anaerobic
Clostridium perfringens	Anaerobic
Clostridium sporogenes	Anaerobic
Clostridium tetani	Anaerobic
Coxiella burnetii	Cell dependent
Enterobacter spp.	Facultative
Erysipelothrix rhusiopathiae	Facultative
Escherichia coli	Facultative
Flavobacterium menigosepticum	Aerobic
Francisella tularensis	Aerobic
Fusobacterium necrophorum	Anaerobic
Klebsiella pneomoniae	Facultative
Legionella pneumophila	Aerobic (Cultured under CO_2)
Leptospira interrogans	Aerobic
Listeria monocytogenes	Facultative
Morganella morganii	Facultative
Mycobacterium tuberculosis	Aerobic
Mycobacterium marinum	Aerobic

SPECIES	CLASSIFICATION
Plesiomonas shigelloides	Facultative
Proteus spp.	Facultative
Pseudomonas aeruginosa	Aerobic
Pseudomonas pseudomalleri	Aerobic
Salmonella typhi	Facultative
Salmonella enteritidis	Facultative
Serratia marcescens	Facultative
Shigella boydii	Facultative
Shigella dysenteriae	Facultative
Shigella flexneri	Facultative
Shigella sonnei	Facultative
Staphylococcus aureus	Facultative
Streptococcus faecalis	Facultative
Vibrio alginolyticus	Facultative
Vibrio cholerae	Facultative
Vibrio parahaemolyticus	Facultative
Vibrio vulnificus	Facultative
Yersina enterocolitica	Facultative

Published by the American Academy of Microbiology, 2000

Chemical Contaminants

Chemical contaminants can be from many sources. Oil can leak from a turbine pump or accumulated oil can escape the casing during low aquifer levels and contaminate the aquifer. Pesticides, fertilizers, or industrial chemicals may infiltrate a well resulting in a contamination problem. If the contaminant is isolated in the well and its immediate environment, a decision may be made to decontaminate the well system. Such a program would be similar to that used for microbial contamination and take into consideration the condition and structure of the well, the biological and mineral involvement, the chemistry required as a cleaner, and the mechanical requirements necessary to deliver the chemistry to the area of the contaminant.

Organic Contaminants

Organic chemicals foreign to a well system or contaminants of that system could be divided into two major groups: the hydrophilic, which have an affinity for water, and the hydrophobic or water insoluble products. This broad system of classification allows us to look at the physical properties of a chemical and how it would be maintained in the well environment. It also allows some selection of the right chemistry for removal. If the product were insoluble we would expect it to reside in areas of low flow to facilitate their accumulation in spaces away from the physical force of the flowing water. In the case of petroleum products we would find traces covering the well structures as a surface coating. Products of low specific gravity, such as the oil, would rise to the higher reaches of the well that may include the gravel pack and formation immediate to the well. All products or contaminants would most likely enter the pore spaces of the well structure with the presence of biofilm and mineral deposits further complicating the contamination.

Inorganic Contaminants

While hydrophobicity and hydrophilicity are characteristics of organic chemicals, inorganic chemicals can also be measured by their ability to be dissolved by water. Inorganic precipitates make up the mineralization of well gravel packs and formation areas ultimately clogging or cementing the pore spaces or flow pathways. Acids are used to dissolve these minerals increasing their water solubility and aiding in the removal when the cleaning solution is pumped from the well. Inorganic toxic chemistries should also respond to similar cleaning both for the removal and/or neutralization as the situation requires. Ultimately the reaction specific to the particular contaminating chemistry must be taken into consideration before any type of cleaning chemistry is used so that any toxic by-products can be accounted for and/or contained.

Contaminants in the Well Environment

While there is a lot that we do not know about microorganisms in the environment, considerable information about how bacteria live and multiply in water systems has been uncovered in the last few years. A tremendous amount of research has gone into the study of biofilms, the natural habitat of bacteria. In water well systems the knowledge of biofilms has led to a much better understanding of bacteria and in general all microstructures in the well. This information has led to the development of better mechanical and chemical practices in the cleaning of wells.

CHAPTER 3 Biofouling Significance in Decontamination

Bacterial growth is responsible for more than 80% of the blockage in wells. Not only are bacteria responsible for biofilm structures that can block flow pathways in the well environment, but the sticky adhesive polymeric material they produce provides an excellent base for mineral accumulation. The matrix that is produced is often a hard, cemented mass capable of extensive blockage of screen and slot openings as well as gravel pack and formation flow paths. In addition, bacterial growth accounts for a major portion of the corrosion process in the well resulting in the accumulation of iron oxides that block flow and cause water quality decline. Bacteria reside in fluid environments, and indeed in wells, either as a free-swimming microorganism, which microbiologists refer to as planktonic, or they are attached and participate in a structure known as biofilm.

Biofilms

The biofilm is actually the natural habitat of the bacteria and consists of the bacteria and the exopolymer they produce. The exopolymer is a polysaccharide polymer that is produced by the bacterium as a means of attaching to a surface, protection from chemical or physical activity, and nutrient capture. There are many different types of exopolymer material, but for the most part they are polysaccharide polymers. Since the polymeric material is insoluble in water it often makes up a large portion of the colloidal matter (suspended solids) found in well water, particularly during a cleaning procedure.

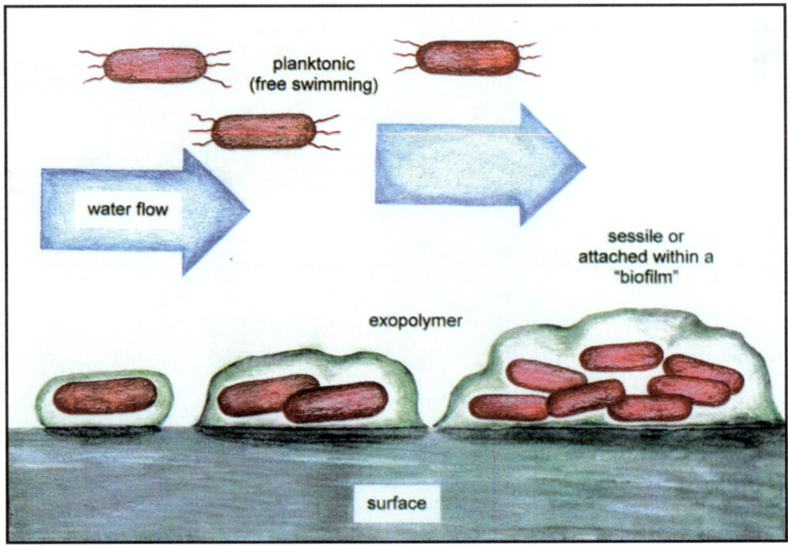

Figure 3.1. Progressive biofilm formation

Figure 3.1 is an attempt to show the development of a biofilm. Note that the free-swimming bacteria land on a surface and by secreting a sticky polymer they are able to stay attached to the surface even in the path of the flowing water. As time passes the bacterium multiply and one bacterium becomes two and so on. This mitosis or cell division is the primary means of reproduction and in due time will produce many bacteria. The bacteria produce biofilms as a place of rest and proliferation. Biofilms are dynamic, that is there are bacteria entering and leaving the biofilm as well as being produced within the biofilm. The features of a biofilm make it an ideal habitat for the bacteria while the presence of the biofilm on the surfaces in a water environment often lead to clogged or blocked flow pathways. A single biofilm may harbor many types of bacteria. Often anaerobic organisms, those requiring no oxygen, and aerobic bacteria, those requiring oxygen to metabolize their food and produce energy, are found in the same biofilm. Of course the anaerobes are usually deeper in the biofilm away from the ingress of oxygen while the aerobes are near the surface or water flow where oxygen can infiltrate.

While it is easy to think of biofilms as just a film or flat thickness, they are actually groupings of mushroom shaped formations surrounded with waterflows that carry necessary oxygen and nutrients. These formations often extend several inches or feet into the water flow. Figure 3.2 is an artistic drawing made from a series of

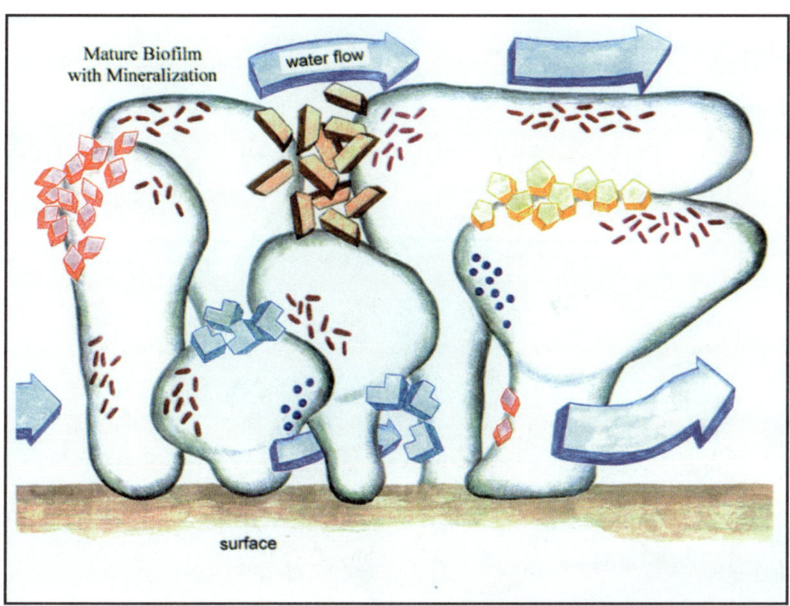

Figure 3.2. Biofilm showing mineral attachment

electron micrographs of a biofilm. Placement of the microorganism is more conceptual in the drawing than shown in the micrograph. Figure 3.3 is an actual photograph taken in a well system. The long extensions of the mushroom-like formations are easily observed.

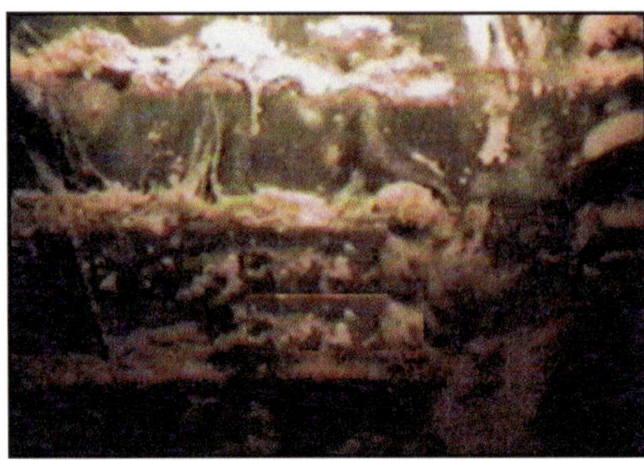

Figure 3.3. Actual photograph of biofilm in a well

Biofilms cover most surfaces within the well environment. The coverage on the surfaces of the geology of the formation, the gravel pack, and of course the screen or other well structure is what ultimately impairs water flow. It is this formation, as shown in Figure 3.4, that complicates the cleaning process. The slimy polymer and entrapped minerals are perfect areas of concealment for pathogens or contaminants of an organic or inorganic nature.

In order to better understand the way bacteria and biofilms are involved in the chemistry and the biology of wells, a brief discussion of the major groups of bacteria that inhabit wells is in order.

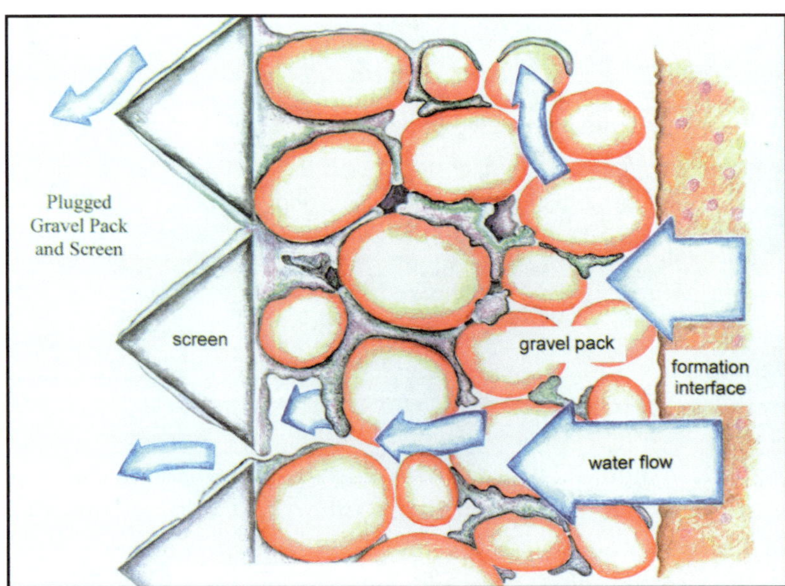

Figure 3.4. Fouled screen/gravel pack

The Common Kinds of Bacteria in Wells

There are thousands of different kinds of bacteria that inhabit well systems; however, those that are known to cause biofouling and result in the blockage, corrosion, or water quality problems are usually in one of the three groups. The slime formers, the iron oxidizers, and the anaerobes (of which the sulfate-reducing bacteria are the most offensive) make up three groups. Other bacteria of note in well work are coliforms that are members of the slime formers and used as indicator organisms for waste water contamination. In addition to the above, the only bacteria not included but occasionally seen in well systems are the branching bacteria that often inhabit the borehole wall in wells that have stood idle for an extended period. These organisms belong to the *Actinomycetes Group* of bacteria and are somewhat like crabgrass. They send their mycelial growth out creeping along the borehole wall and eventually seal off flow to the well. While some are facultative, most are aerobic so growth is more often seen in shallow wells where aquifers maintain a certain level of oxygen. Wells that have set idle for several years should be biologically tested and, if necessary, chemically cleaned before putting back into production.

Slime Formers

The slime formers are the largest group of bacteria and include members from just about all the families of bacteria; that is, many of the soil bacteria are represented as are the coliforms and most of the environmentally active organisms such as the *Pseudomonas*. These organisms are similar in that they tend to produce considerable exopolymer or slime. In the well environment we see them on almost all surfaces and certainly they are known to inhabit or impact the gravel pack and aquifer formation, particularly the areas just outside the borehole wall (see Figure 3.5). These are the primary populations that provide extensive biofilm plugging in the aquifer or zones around the well. The slime formers include just about all types but are primarily aerobic or facultative anaerobes, that is they are capable of existing in both an aerobic and an anaerobic environment.

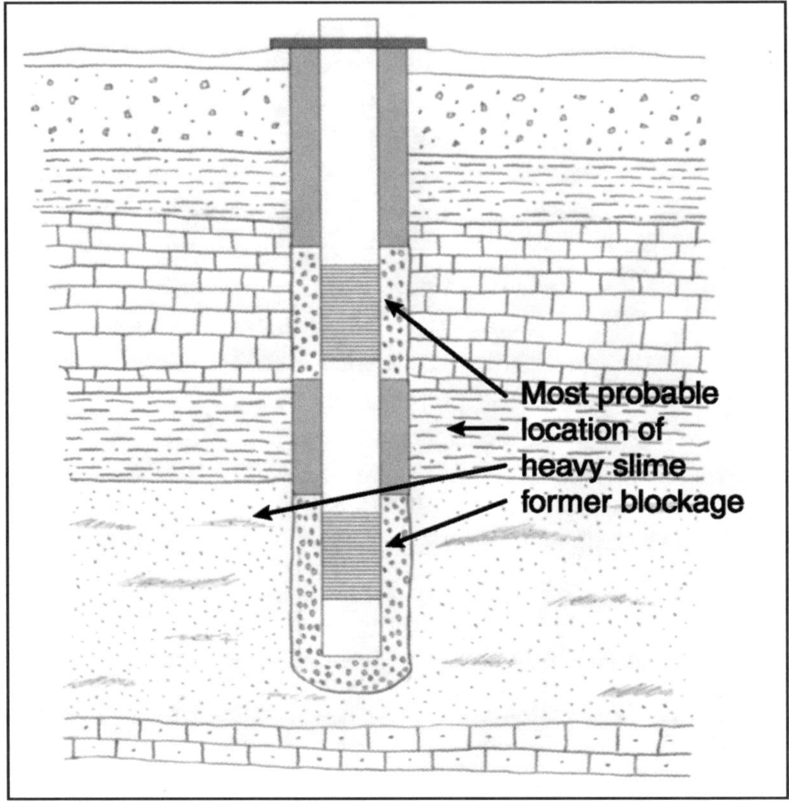

Figure 3.5. Probable locations of bacterial blockage

The importance of this group, other than that they include so many of the bacteria, is the way they respond to change. Increases in flow, nutrients, or oxygen can bring about a marked increase in biofilm production. Use of certain chemicals also evokes more polysaccharide production. These bacteria provide much of the regrowth notcd after chlorinating procedures (Characklis and Marshall, 1990). Certain pairs or groupings of these bacteria are known to produce both a greater volume and a denser biofilm when present together than when only one of the populations is present

(Molin, 2000). Because of the many types of metabolism and biochemistry present in the various members, the slime former group is capable of proliferating on many types of nutrients and/or environmental conditions. Typical members of this group other than the coliforms are the *Pseudomonas, the Flavobacter, Acineto-bacter, Aeromonas,* etc.

Iron Oxidizers

Many of the slime formers are capable of oxidizing iron because of the almost universality of their members. In this book, however, I will refer to the typical sheath-forming bacteria as the iron oxidizers primarily responsible for screen blockage or blockage due to heavy accumulation on the borehole wall. These bacteria produce the slimy oxidized iron, iron oxyhydroxide, from the oxidation of ferrous iron and store it in or on long stalks or sheaths which add to the blocking matrix. The oxidation of manganese usually takes place in the same manner. A more realistic way of classifying these organisms would be to call them iron accumulating bacteria. In our laboratory we usually classify these as those identifiable with the use of the light microscope and characterized by the formation of some type of structure. They are not easily cultured in the laboratory but are observed during routine microscopic examination of the sample.

Some of the iron oxidizing bacteria carry out their basic metabolism with energy achieved from the oxidation of ferrous iron according to the following formula.

Table 3.1. Energy reaction for bacterial iron oxidation

IRON OXIDIZING BACTERIA
Energy for growth from the reaction:
$4Fe^{2+} + 4H^{+} + O_2 \rightarrow 4Fe^{3+} + 2H_2O + \text{energy}$
Which leads to this reaction:
$Fe^{3} + H_2O \rightarrow FeO(OH)$
Ferric Oxyhydroxide is very insoluble and plugs flow pathways

As noted in Table 3.1, once the ferric ion is produced iron oxyhydroxide is then formed from its reaction with water. This material is a dense slick substance capable of considerable plugging potential. In addition the conversion reaction is very limited energy wise and therefore considerable oxidation must take place in order to supply

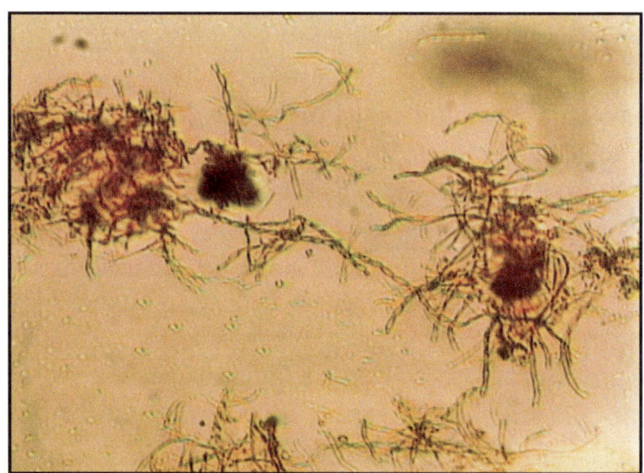

Figure 3.6. Microphotograph of the iron oxidizing bacteria

the population with sufficient energy to sustain growth. Figure 3.6 is a microphotograph of a *Gallionella ferruginea,* probably the most common iron oxidizer of this group. The twisted stalk is characteristic of this specie, however in general most microorganisms in the group produce some type of stalk or distinct formation. The *Gallionella* actually fill the stalk with the oxidized iron and release the stalk into the flow producing more stalks as additional iron is oxidized. Other members such as *Crenothrix* and *Leptothrix* deposit iron and manganese oxides on the exterior of the sheath-like structures.

The importance of these organisms is the matrix made up of the sheaths, exopolymer, and the iron oxide that results in a very dense biofouling or blockage. The tangled matrix as observed in Figure 3.6 causes bridging of screen openings resulting in rapid diverting of water flow. As the bridging progresses, infiltration of the gravel pack follows and eventually severely limits well production. At first the deposit is soft and viscous, adhering to screens and casing. As time passes the iron oxyhydroxide dehydrates and the deposit becomes extremely hard and difficult to remove. Most often the deposits are found up close or on the screen or slot openings in the casing, but as indicated as the biofilm matures this population will extend into the gravel pack.

In open borehole wells the borehole wall often becomes impacted with large populations of these bacteria; however, in these wells we usually see more mobility and the organisms are often seen in the distribution system.

It has been our observation that while all types of iron oxidizers are often found in the distribution systems, it is the *Gallionella* that are more often seen in abundance. These are the bacteria (or one of the others of the stalk forming group), or more particularly the stuffed or encrusted stalks that constitute the so-called "organic iron," that are often reported in distribution water. These bacteria are responsible for filter plugging and small line blockage often seen in industrial systems. It is also interesting to note that in most systems where *Gallionella* had infiltrated a distribution system the organism died off or simply disappeared once the well source for the water was cleaned. No subsequent cleaning of the distribution was necessary.

Anaerobic Bacteria

The anaerobes are those bacteria that do not require oxygen to live. In fact many anaerobic bacteria die in the presence of oxygen. These microorganisms are found deep in the well or gravel pack where oxygen is very limited or not present at all. They produce a fairly dense biofilm in order to blank out the oxygen. This could lead to severe blockage in the lower areas. In addition most anaerobes produce gaseous products as the end result of their metabolism. These gaseous products are responsible not only for pH changes or the resulting corrosion, but often lead to severe water quality loss in the distributed water. The most often encountered members of the group are the sulfate-reducing bacteria (SRB).

Sulfate-Reducing Bacteria

SRBs are a specialized group of anaerobic bacteria that are known for the hydrogen sulfide gas or rotten egg odor they produce. The H_2S is formed by the reduction of sulfate and is also a major source of mineral sulfides both in wells and aquifer formations. The sulfate-reducers are also responsible for the oxidation of products of the anaerobic fermentative bacteria (fatty acids, alcohols, aromatic acids and hydrogen). Through this oxidation they supply the final step in the decomposition of organic material in the anoxic areas of the well. Therefore, in the well bottom or sump where water flow is often restricted and organic matter accumulates, large amounts of gaseous (acidic) products are produced.

The H_2S gas often results in corrosion in the upper reaches of the well as the strong acid gas is forced upward. Contact between the steel casing and column pipe, with the acid formed, results in corrosion of the steel surface and formation of iron sulfide. Later the iron sulfide is oxidized in the oxygen rich environment and the sulfide is oxidized to pure sulfur. This formation produces large raised blisters or tubercles on the surface which when scrapped or opened are filled with yellow crystalline sulfur (Figure 3.7).

The tubercle formation results in destruction of this part of the well structure, often with complete penetration of the piping, while the initial growth of the bacteria results in loss of the bottom segment of casing or screen. Sulfate-reducing bacteria are often the first indication that the well has a significant anaerobic population.

Figure 3.7. Sulfur containing tubercles

Coliforms

These bacteria are really not a separate group but actually belong to the slime formers because they produce considerable exopolymer slime. The coliforms are important because they are more regulated than any other organisms in wells. They are usually considered by regulatory agencies and state health departments as contaminants. In actuality they are only indicators of possible contamination.

Coliforms are facultative, that is they can live in the presence of oxygen or in anaerobic conditions. This ability provides the coliform with a strong defense mechanism. When the free flowing zones, the aerobic zones, of a well are flooded with a chlorine solution, the coliforms are able to survive deep in the well by inhabiting the anaerobic biofilm. This biofilm is more dense and is not as easily penetrated or reached as the aerobic biofilms in the casing, screen, and other free flowing areas (Figure 1.1). The hydrogen sulfide gas produced by sulfate-reducers is a signal that housekeeping in the well has become lax and that debris has built up sufficiently to produce a strong anaerobic zone capable of harboring coliform organisms.

Microbial Induced Mineralization

In the text above we discussed the iron oxides that were produced by the action of the iron oxidizing bacteria but almost more universal are the other minerals that accumulate in the slimy biofilms in the gravel pack and formation. The mineralization of a biofilm, or more graphically the slime filled spaces between gravel or the openings in casings and screen, takes place gradually as small crystals are caught in the adhesive mass and gradually grow until they consume most of the space. This mineralization process effectively cements large areas of the gravel pack and the formation blocking flow pathways. The process is probably one of the more debilitating natural phenomena that takes place in the well.

Both laboratory research and observation of operating well chemistry have shown that most of the time well water is fairly evenly balanced and may operate many years without significant mineral blockage. Once a biofilm develops in the well, however, mineral formation takes place and significant blockage occurs. Figure 3.8 shows a gravel pack containing a large deposit of calcite, calcium carbonate, formed in one area. This formation was actually repeated throughout the gravel pack. The calcite at this time had not completely filled all the areas between the gravel. The calcite mineralization formed initially from the first crystals entrapped and then grew into a sizable mass. Eventually it would have grown reaching out to other deposits and in effect cementing the whole gravel pack.

Areas of Contaminant Accumulation

The process of biofouling, mineralization, and/or a combination of both together with the intricate flow pathways that make up a well environment, offer considerable area for contaminant accumulation. Pathogenic organisms can pass through the well or they can become part of the resident biofilm. Chemical toxins can be absorbed into the polymer matrix or adsorbed on the mineral surfaces or, as the free-swimming microorganisms, pass through the well environment. Chemical contaminants, particularly the inorganics, can become part of the mineral formation especially the

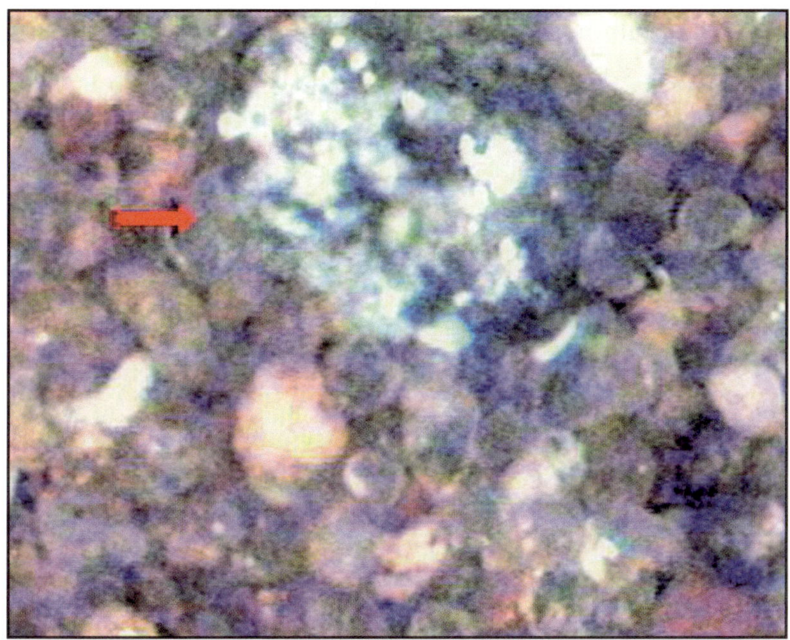

Figure 3.8. Carbonate mineral accumulation in gravel pack

newly formed or forming mineral deposits. Chemical contaminants, such as hydrocarbons, oils, and volatile and nonvolatile organics, can not only absorb on the mineral structures but can also become part of the biomass particularly as a food source (Spath and Flemming, 1998). Since the surface area involved in a well and the participating pore spaces is enormous, the areas of possible fouling or contamination are great. More important than the expanse of the area is the type of contaminant and how it is construed within the well.

Chemical Activity Against Biofilm

Most of the structural blockage in wells is mineral and responds to acid cleaning. However, since many of the complications of well contamination will be the result of the presence of biofilm, some knowledge of the cleaning or removing of biofilms is important. Eighty-five percent of most biofilms consists of the extra cellular polymeric substance or EPS, usually a mixture of protein, nucleic acid, lipids, and polysaccharides. The polysaccharide is considered to be the most abundant, but this is not always the case (Flemming and Wingender, 2001). This matrix of biopolymers exhibits considerable resiliency against chemical cleaning. The following data is a review of the various chemicals that have been and are still used for cleaning and/or removing biofilms. The information has been collected from both laboratory and field trials. References are given where available and appropriate.

Mineral Acids

Generally these acids produce very little damage to biofilms with the possible exception of dewatering due to acid concentration and the mechanical effect achieved when acid is applied. There are, however, some physical disruptions when the acid attacks biofilms heavily impregnated with carbonate formation.

Organic Acids

Two organic acids, hydroxyacetic and acetic acid, are thought to be somewhat effective in biofilm dissolution. Usually they have to be at a concentration between 5 and 10% to have much effect. They are often used in conjunction with mineral acids that control or dissolve the mineral accumulation.

There are two serious drawbacks to their use. First, most organic acids produce only partially soluble mineral salts that are often left in the formation. Secondly, they all have a high carbon content which is an excellent food source for bacteria.

Biodispersants

These polymeric cleaning products have shown excellent activity during the acid cleaning of well systems. They appear to impart considerably more cleaning activity when used with the acids for the removal of the combined mineral biofouling than when the acids are used alone. Some of the improved cleaning effect is, of course, directed against the dissolved minerals and their removal following the cleaning activity. Investigated gravel packs heavily biofilm-impacted were cleaned much more efficiently with acid blends using specific biodispersants than with acid cleaning alone (Schnieders, M. and Adkinson, 2000). Recent work by Flemming and Wingender (2001) in which they discuss the role of calcium, copper, ferrous, and ferric iron in the formation of the extra cellular polymeric material sheds considerable light on the reasons for this apparent effectiveness. Biofilm of well origin most often contains a ready abundance of these cations resulting in a biofilm formation that may be predominately dependent on the electrostatic interaction established with these particular cations. Tested biodispersants such as NuWell-310, QC-21, and UNICID Catalyst have strong dispersant and chelation-like properties against these metals and may explain the exceptional ability of these products in disrupting the heavy biofilm characteristic of a plugged or biofouled well.

Surfactants

Surface-active agents are added to most or certainly many cleaning solutions and this has been true of most biofilm cleaning preparations. Our observation during well cleanings and also in our laboratory experiments is that the surfactant does not contribute to dissolution or removal of the biofilm. This we feel is particularly true in biofilms of deep well origin. The work sited earlier of Flemming and Wingender also showed that surfactants were particularly ineffective where calcium provided the main binding force of the polysaccharide strands. They suggest that the surfactant activity is directed more toward dispersion and that the binding force of the exopolymer strands is more cohesive as a result of the electrostatic interactions. Cleaning procedures do benefit from surfactant addition however, more because of increased penetration of the mineral blockage and the formation than from its effect on biofilm.

Chlorine or Other Oxidizers

Chlorine, in its different forms, and other oxidizers, such as hydrogen peroxide and ozone, are the primary chemicals used as disinfectants to destroy pathogenic organisms and other bacteria accumulating in wells. It is assumed it will "burn up" or oxidize the biomass found in well systems because of its strong oxidizing power. To some extent this is true when a strong concentration of the chemistry can come into direct contact with the polymeric material (polysaccharides, nucleic acids, proteins, etc.) and shear force or flow is sufficient to wash away the oxidized debris and place fresh oxidizing chemistry on the remaining biofilm and the action can be repeated until the biofilm is destroyed. Unfortunately, while the chlorine dose is often extremely strong, the removal or scrubbing action is insufficient. In addition, oxidizing chemistry actually changes the biopolymer into a less soluble material. In the case of polysaccharides, oxidizing agents will form a chemical gum that is often less water soluble and more difficult to remove (Brewster and McEwen, 1963 and Characklis, 1980). See Figure 3.9.

Well systems chlorinated using pH control at a pH of 6.5, with chlorine levels held between 50 and 200 mg/l, have been more successfully decontaminated. This may be explained if you consider the excessive oxidation rates with the high levels of chlorine most often applied unsuccessfully. In poorly maintained wells, or those with high levels of biofouling, the high chlorine (oxidizer) level may build a very tough barrier which in effect protects those bacteria in the deeper recesses of the biofilm.

Enzymes

Since biofilm appear as a ready source of carbon, the idea of using particular organisms for its destruction or the enzymes they produce as a means of dispersing or dissolving the biofouling appeared to be a natural solution. "Evolution has obviously not generated organisms that readily degrade these matrices, otherwise such organisms would have one of the largest carbon sources one can think of" (Flemming and Wingender, 2001). This would indicate enzymes are not a good choice and have been very ineffective when used for

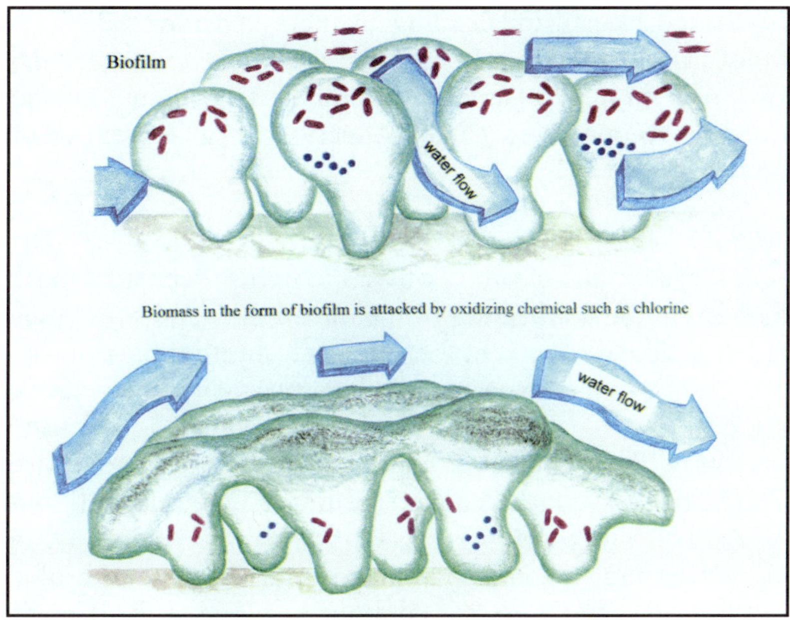

Figure 3.9. Chlorine attack on biofilm

biofilm removal. This is consistent with work in our laboratory and that reported by Klahre and Flemming (2000).

Chemical/Mechanical Activity for Biofilm Removal

Since biofilms are most always accompanied (in the well environment) with mineral formation involved within its architecture, the most effective cleaning chemistry must contain some mineral acid. The concentration of acid utilized should be dependent on the potential for mineral involvement and in particular the presence of calcium carbonate. Mineral acids have only momentary disruptive power against biofilm (for instance as the carbonate molecule is broken apart); therefore, some additive has to be used to improve biofilm solubility and aid in the removal from the well during pump out.

Biodispersants, as noted earlier in text, are designed to be used with the acids to specifically aid in biofilm solubility (Schnieders J., 1996) and suspension leading to increased removal. Use of a blend of acid and polymer chemistry will provide cleaning or removal of both mineral and biofilm blockage without oxidation of the organics (biopolymers) and production of subsequent insoluble materials.

"Without elbow grease most things don't get cleaned." The old proverb is true in the cleaning of well systems. Persistent mechanical activity in the application of the cleaning chemistry is required to be successful in the removal of heavy biofouling from a well system. While many types of applications have been developed over the years, the consistently effective method is the use of a double surge block. However, jetting tools may be more adaptable to open borehole wells since damage to the borehole wall must be avoided. In cleaning well systems with gravel pack, screens, and/or perforated casing, the movement of water as the surge block is moved up and down creates a continuous swipe at the deposit and effectively moves the dissolved material out of the reaction pathway. Laboratory and field studies have shown that the water movement extending out into the gravel pack produces movement in the formation water at a considerable distance from the borehole. This is particularly true of wire wrap screen construction because of the total open area and the ability of the cleaning fluid to move horizontal to the well in a larger volume.

CHAPTER 4 Biological Activity and the Cause of Specific Well Problems

In the previous chapter we reviewed the making of a biofilm (Figure 3.1) and late in the chapter the activity of chlorine and other oxidizers against the biofilm structure (Figure 3.9). Several other facts concerning the metabolism and growth of bacteria are also important and should help to achieve a better overall picture of bacteria and biofilms in the well environment.

The Number of Bacteria in Biofilm

First, the ratio of free-swimming (planktonic) bacteria to those in biofilms (sessile) is 1 x 10^6 or one to a million. Considering that we are only able to culture less than 1.0% of the bacteria in water (Finchel and others, 1998) you can see that the potential number of organisms in the biofilm is quite enormous. These figures can also be used to give some perspective to the heterotrophic plate counts (HPCs) which actually only account for free-swimming aerobic organisms. Their use, however, can be helpful if regular samples are taken and changes are viewed over a substantial period. For instance, if plate counts were run only quarterly it would be difficult to track the normal cycles of a bacterial population. Bacterial populations cycle up and down over time. On the other hand, if weekly plate counts were run and changes in population numbers were observed over time, it would be more logical to expect to derive good information concerning the changes in population size or the biological load placed on a well system.

The Exponential Growth of Bacteria

A second characteristic, the generation time for most types of bacteria, varies from 20 minutes to several hours. Figure 4.1 shows the exponential growth of bacteria with a 20-minute generation cycle. In a little over three hours the bacterium has multiplied 1,000 fold. This, of course, is under an ideal situation where sufficient

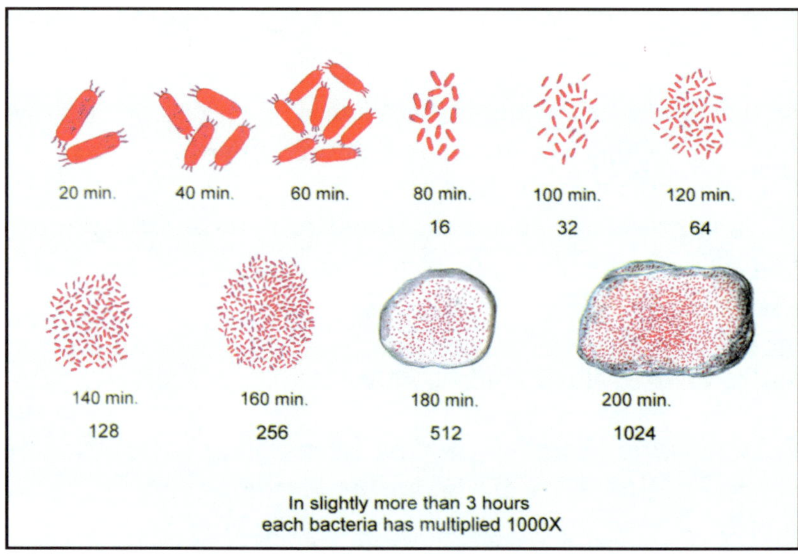

Figure 4.1. Exponential growth of bacteria

food was available and no deaths occurred. Nevertheless the growth potential is extremely high and can be used to explain some of the phenomena we see everyday in well operation.

I often ask the question, if a bacterium were placed in the middle of a room and that room represented a path for water flow, what would be the sequence of events if the bacterium received its needs for nourishment and waste removal? After a great deal of time we would begin to see a growth and a little later that growth would appear to be expanding much faster. In fact, growth would continue rapidly until half the room was full. Now, water flow is directed around this blockage and for the most part no reduction in flow has been observed. What would happen in the next 20 minutes? We would see a complete blockage of all flow. In other words, even though the mass of bacterial growth was doubling every 20 minutes it was so small it made very little impression on the water flow. Once it had achieved "critical mass," so to speak, it was able to block flow completely very quickly. What well consultant or contractor has not received a call from a client who was complaining about production loss from a well ten or fifteen years old. *"It could not possibly be bacteria because the loss just started to develop!"* Production loss won't be 100% and usually does not happen overnight, but it can appear in a relatively short time due to the bacterial growth over an extended period.

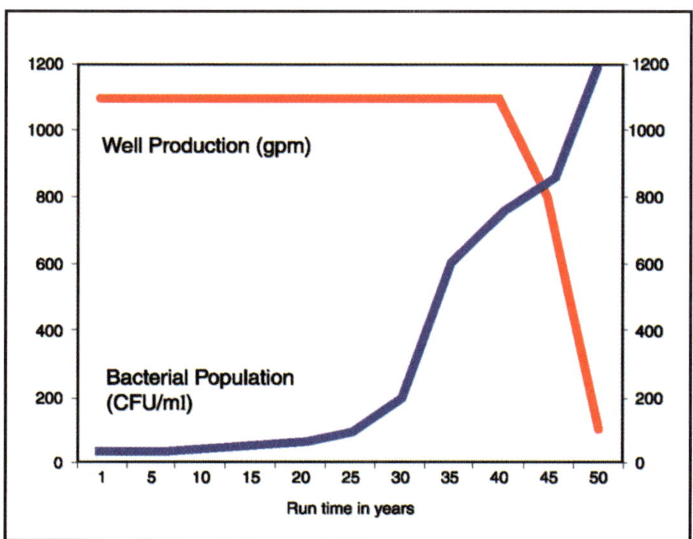

Figure 4.2. Bacterial growth vs production – 50 year well

Figure 4.2 shows a well in California that was 50 years of age when the data was reviewed. It shows a graph depicting the growth of bacteria in the well. The mass of bacteria gradually increased for 50 years. At age 45 the well began to experience loss of production and by age 50 had lost more than 95% of its original capacity. It had been decided at age 45 not to rehabilitate the well because of its age. Unfortunately in the last five years water needs had escalated and the city was in need of its former production. Analysis showed the well was severely biofouled with little probability of mineral plugging. Loss of production was due to the gradual increase in biogrowth that could have been prevented if the well had been periodically cleaned. If someone had removed the bacterial growth in our make-believe room when it first appeared, the room would have never been filled. Preventative maintenance, which could include some chemical cleaning when pump maintenance is performed, would go a long way in keeping a well free of blockage.

Growth of Bacteria in the Water Column

A third phenomenon we often see in well systems has to do with the bacterial growth throughout the water column. In most well systems the column of water, while agitated during pumping, spends considerable time in a quiet non-mixing state. During these quiet periods bacterial life thrives at all levels. In Figure 4.3 the top part of the water column shows the presence of aerobic bacteria. This community of bacteria uses molecular oxygen as the final electron acceptor. They are able to do this because of catalase, an enzyme they produce that breaks down hydrogen peroxide into water and molecular oxygen (when oxygen is dissolved in water, hydrogen peroxide is formed). Since hydrogen peroxide is a strong oxidant, the catalase not only provides oxygen as a final electron acceptor but protects the cell from damage.

Further down in the column the oxygen becomes less concentrated and those bacteria such as the iron oxidizer *Gallionella* live in this zone that divides the area of oxygenated water from the anoxic zones near the bottom. *Gallionella* obtain their energy from the oxidation of ferrous iron which is present primarily in the anoxic zone, but since it is an aerobic organism it requires oxygen.

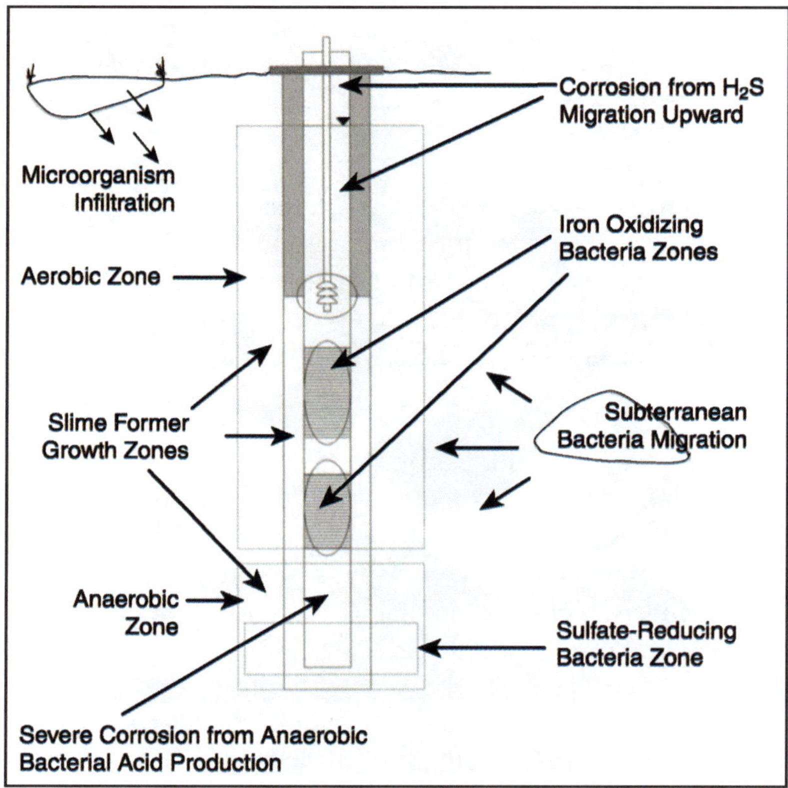

Figure 4.3. Water column

These apparent contradictive needs require the bacteria to reside in or near the interface of the aerobic and anaerobic zones.

As the water column becomes less oxygenated the water turns more anoxic and the anaerobic bacteria are then the dominant population present. The anaerobes are comprised of many types of bacteria. One of the major populations present are the sulfate-reducing bacteria or SRBs. These organisms however are not fermenters, that is they cannot completely metabolize carbohydrates to carbon dioxide and water. They are dependent on associated fermentative bacteria to oxidize partially the carbohydrates to simple organic acids and hydrogen. These organic acids and hydrogen are then coupled with sulfate-reduction in order to provide energy for growth. The end result is the production of hydrogen sulfide gas. By preventing the build up of fermentative by-products (organic acids and hydrogen) the SRBs are said to achieve a symbiotic relationship with the fermentative bacteria (Fenchel and others, 1998).

While density gradients in the water column prevent the movement of oxygen from the upper region downward, it does not prevent particulate matter from settling downward. As aerobic growth dies off bio debris settles toward the anaerobic zone. This accumulation of organic substance becomes a food source for the anaerobic populations.

What does the column mean to good well operation? Starting at the top any incident or parameter which would allow aeration of the water or infiltration of aerated water could result in an excess growth of aerobic bacteria. Overgrowth and die off would result in a considerable increase in available food for anaerobic bacteria. The fermentative and sulfate-reducing activity of the anaerobic bacteria would result in the release of hydrogen sulfide gas. This gas along with excess organic acids and hydrogen (from the fermenting anaerobes) would produce a very acidic or corrosive environment. Severe damage could occur to the well structure. As the hydrogen sulfide gas moves up the well column the acidity would result in generalized corrosion and the hydrogen sulfide dissolved in the water would cause considerable complaints of taste and odor problems (rotten egg odor). Gas that reached the upper well above the water level would dissolve in the water vapor condensed on the column pipe and drop pipe resulting in corrosion of these surfaces with iron sulfide formed. Later oxidation of the sulfide would result in tubercles or "blisters" filled with elemental sulfur (see Figure 3.7).

In addition to taste and odor problems, the dissolved hydrogen sulfide often encourages the growth of sulfur oxidizing bacteria such as *Thiothrix.* These bacteria oxidize the sulfide to sulfate as an energy source and often appear as black spots in the distribution water (see Figure 4.4). Cleaning of the distribution system is usually not required as the *Thiothrix* disappears once the well or source of sulfides (sulfate-reducing bacteria) has been cleaned.

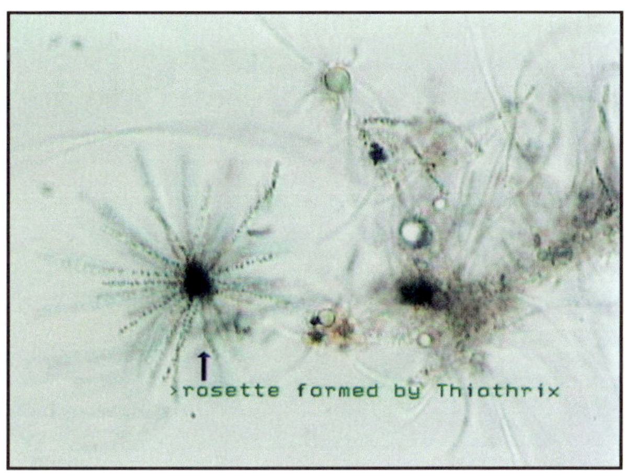

Figure 4.4. ***Thiothrix*** **bacteria**

In 1990, a small community in Illinois had a well sample analyzed and we found numerous sulfate-reducing bacteria. During a conversation with the contractor in which I was trying to emphasize the necessity of cleaning the well bottom of any debris and growth, I was told that the well had been constructed with a 15 foot sump and there was no flow in that area of the well. I was also told the story of how the well was cleaned every two years and that while the water was excellent for a few months the terrible odor and taste always reappeared. In consultation with the well superintendent and the contractor, we decided after cleaning to cement the well sump to prevent any accumulation of debris in that section of the well. The well has had excellent water ever since.

Many wells are constructed with sumps (dead areas) to collect sand or particulate matter that might enter the well. Unfortunately where organic constituents of the water are high and aerobic organisms are overactive in the water column, these sumps become ideal environments for sulfate-reducing bacteria. We have used similar recommendations when the incidence of hydrogen sulfide production has persisted in a well's history. Each time and in many areas of the country we have succeeded in solving a long-term problem.

CHAPTER 5 Coliform and Pathogens in Wells

Coliforms exist in wells in certain environments and under certain conditions that make them more difficult to remove. As such, their removal is a good source of information on the methods that could be applicable in the removal of pathogens or other microbial contaminants of a well system. Their usual low populations and difficulty in removal make them a good model of a limited infiltration or recent entrance of a non-typical resident into the well environment.

Coliforms

Coliforms are bacteria used as indicators of waste or fecal contamination in drinking water supplies. These bacteria were written into regulations because they are relatively easy to test for and because it was believed at the time (early 20th Century) that they only reproduced in the gut of warm-blooded animals, including man. It was reasoned that if coliforms were present in a water supply, there was some contamination from a wastewater source. Since very little work was done which involved searches for "coliforms" in environments other than potable water and its treatment and distribution systems, only a few questioned the theory of coliforms and contamination.

As time and science progressed, we came to realize that coliforms, like so many bacteria, are quite capable of living outside the gut of warm-blooded animals and are often found in the environment (Hursts and others, 2002). "The coliform group is a heterogenous conglomerate of microorganisms, including forms native to mammalian gastrointestinal tracts, as well as a number of exclusively soil forms" (Applegate and Erkenbrecher, 1989).

Coliforms are not a particular species of bacteria. They are a group of bacteria made up of several different genera and species that conform to certain growth requirements. These requirements are present in all the separate species to different degrees and some respond to traditional laboratory analysis quite readily while others do not. Identification differences exist in traditional laboratory tests and result in subtle differences in the organisms enumerated (Clark and Pagel, 1977).

The Colilert test was developed more recently (within the last 15 years) to detect the presence of coliform bacteria and the simultaneous determination of the presence or absence of *E. coli.* It specifically tests for the presence of B-galactosidase and B-glucoronidase, enzymes specific (so it was believed) to the coliform bacteria and *E. coli* respectively. With this development we were able to test for one bacterium in 100 milliliters of water. Since the Colilert is a very sensitive and easy to run test it has brought about changes in our approach to laboratory results. Performed more often (as more testing is required by law or regulations and more systems are coming under these regulations), the more sensitive Colilert is branding more and more water supplies "contaminated". Since this triggers action and reaction, numerous politicians and regulators and even a few researchers are picking up the standard and warning America of the "contaminated" condition of our water supplies. Nowhere has this been more devastating to the confidence in our water supply and to the escalation of costs for correction than in the groundwater industry. All this for an organism or organisms that are at best only indicators and not even well defined as to their position in the bacterial community.

A Brief Expansion of the Coliform Discussion

Coliform bacteria belong to the family *Enterobacteriaceae* and include *Escherichia coli*, as well as various members of the genera *Enterobacter*, *Klebsiella*, *Citrobacter*, and others. The bacteria of this group have been used as indicators of possible fecal contamination or water pollution from wastewater because they can originate from the intestinal tracts of homeothermic animals. Because of this they are of sanitary importance; however, other bacteria dominate that type of bacterial community (Geldreich, 1978). In addition, coliforms have been proven to originate from nonenteric environments such as waste from the wood industry (Wun, Walker, and Litsky, 1976), algal-mat communities in pristine streams (McFeters, Bissonnette, and others, 1974), and even from biofilms within drinking water distribution systems (LeChevallier, 1990). The ability of these organisms to grow in natural waters, and the fact that they are not even the dominant organism from the intestinal tract of humans or other warm blooded animals, provides serious doubt to their use as a general indicator of bacterial water quality.

A subset of the total coliforms group, of which *E. coli* is a member, is known as fecal coliforms or more correctly the thermotolerant coliforms. These organisms conform to all the criteria of the total coliforms group with the added requirement that they grow and ferment lactose with gas and acid production at 45.5 ± 0.2°C. Pourcher and others (1991), found excellent positive correlation with fecal contamination from warm-blooded animals. Further in their favor as a more specific indicator of fecal contamination is the physiological basis of this higher temperature requirement. The requirement has been described by Clark in 1990 as a thermotolerant adaptation of proteins to their ability at the temperature found in the enteric tracts of animals, which is more constant and higher than temperature found in most aquatic environments. Even though some thermotolerant coliforms have been isolated from environmental samples in the apparent absence of fecal pollution, the incidence is more rare, leaving this group of organisms far more indicative of true pollution than reliance on the total coliforms group as a whole.

While reliance on the total coliforms test to determine the need for disinfection has provided good results and in general served the water production industry well over many years, there are new problems arising as various regulations run head-to-head. The repeated disinfection of systems and the installation of very high priced treatment plants should have a more sound basis than the mere possibility of contamination, particularly when the indicator continues to have little scientific basis for its selection. In Europe *E. coli* has been used effectively as a water quality indicator for some time and more recently has been incorporated into U.S. regulations as an indicator of fecal contamination. The total coliforms test continues to be that which dictates our water quality standard and enforcement or disinfection activity, particularly in the groundwater industry. A positive coliforms test is left open to regulators and users to interpret as a true contamination, usually implied and treated as fecal pollution.

More politicians and regulators are pointing to the rise in positive coliforms detection as an indicator of widespread water contamination. This action has resulted in unprecedented pressure on the well professional to rechlorinate and super chlorinate these labeled “contaminated wells.”

If coliforms are to continue as the indicator organism then they should be used within the scope of what they are, an indicator of possible contamination and not a contaminant. We should instead refine or expand the theory behind the use. The use of a pass/fail test could continue as the first line of detection but should be followed by enumeration tests if a positive is recorded. Our laboratory and a number of contractors, health department professionals, and other laboratories have recorded considerable differences in the number of coliform bacteria found in various labeled contaminated wells. Obviously, high counts could signal a more probable contamination of the well by a wastewater source. Low counts could signal the presence of coliforms species better classified as soil organisms and not representative of a waste contaminating source which have set up a low-grade residence in the well ecosystem. Perhaps an even better observation would be to culture for thermotolerant coliforms. Their presence would be more specific for a wastewater contamination.

Coliforms in the Well Environment

Realizing that the contractors and the county health professionals had a problem, our laboratory at the request of the Michigan Department of Environmental Quality (DEQ), began to study the problem of coliforms in wells. The study, while it did not result in an absolute method for the removal of any residual coliforms, led to a better understanding of how coliforms reside in well systems and what steps can be used to increase the effectiveness of chlorination and the over all process of disinfection.

First, recall some facts about bacteria. Bacteria are either free swimming (planktonic) or they are attached (sessile). The free-swimming bacteria are those that are sampled when we collect samples for coliform testing. The approximate ratio is one to one million (1×10^6). For every one free-swimmer we have a million bacteria attached to surfaces. Organisms grow and reproduce for the most part attached to a surface. This attachment is in the form of a biofilm (see Chapter 3).

Biofilm react to changing conditions just as we react to changes in our environment. If the flow of water is increased so that it moves over the surface of the biofilm consistently but faster, as is the case of a well being pumped, the biofilm tightens allowing fewer bacteria to leave the habitat. However if the well is left idle, biofilm expands with additional water absorption and releases more bacteria into the water flow. When our laboratory attempts to characterize the biofouling in a well system we analyze two separate samples. One collected after the well has set idle overnight and one after the well has been pumped for three hours. Bacterial counts of these samples usually show counts in the first sample five to eight times higher than those from the sample collected after extended pumping.

If chlorine or some other oxidizing chemistry contacts the biofilm it shrinks and becomes hardened. The resulting oxidized material (surfaces of the biofilm) is actually less penetratable and is more protective of the bacteria. This collapse, or shrinking of the biofilm, allows more water flow for a period of time as the flow areas between the gravel are more open. Many well owners observe better flow following a super chlorination only to observe this gain "lost" as the biofilm again begins to grow and move closer together to block the flow pathways.

Biofilms contain some bacteria which are anaerobic; that is they can live without oxygen. It's these bacteria that grow when the top layers are sealed, and once the biofilm is again active and is more opened to oxygen from the water, the aerobic (those using oxygen) bacteria grow and expand the biofilm. All the organic debris left from the oxidation caused by the chlorine is then available as a food supply for these bacteria (Van der Kooij, Hijnen, and Kriuthof, 1989). Bacteria, as mentioned above, are either aerobic, using oxygen, or anaerobic, not requiring oxygen to live. Actually there are others that fall in between; the microaerophilic, those that need a very small amount of oxygen and the facultative, those that can live either aerobically or anaerobically.

The Michigan Study

The Michigan study had the objective of quickly reducing the number of rechlorinations before a well could be accepted. To achieve this a two-fold approach was made. We coordinated our laboratory work, which was designed to study how coliforms lived in wells, with observations of field applications of chlorine by contractors throughout the state. A series of workshops were held in which over 300 well contractors discussed their failures and successes of, to be specific, their "good" and "failed methods" of chlorination. County and state health department personnel were also invited for their input.

Prior to the workshop we had determined in our laboratory that a dosage of between 50 and 500 mg/l chlorine appeared to be the most efficient level for disinfection (Schnieders J, 2001). Higher doses had consistently shown residual coliforms in pumped water from our laboratory test units (Figure 5.1).

Figure 5.1. Coliform testing

A concentration range of 50 mg/l to 200 mg/l was the most consistent effective level. At the 200 mg/l level there was sufficient chlorine to provide a strong concentration on the second day even in wells with a high chlorine demand from heavy bacterial presence. All disinfection procedures were allowed to set overnight before pump out.

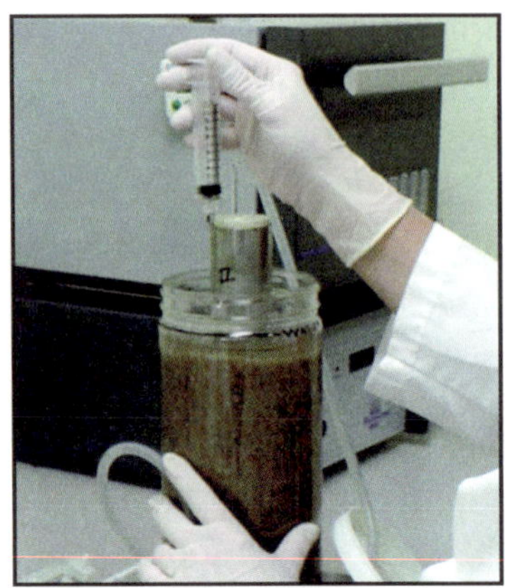

Figure 5.2. Probe allows discreet sampling in the wall

All levels above 500 mg/l chlorine tested positive for coliforms when checked after one week. It was believed that the strong oxidative effect of the high chlorine concentration temporarily sealed off biofilm and in effect protected the bacteria. Needle probes into the mock well gravel pack revealed pockets of bacteria, almost always anaerobic, that contained coliform bacteria. Subsequent testing of areas of free water flow and areas in the upper zones more subject to aerobic growth in the gravel pack or formation showed no coliforms once the well had been chlorinated (see Figure 5.2). From these findings we hypothesized that disinfection requires more than the addition of chlorine or even the pH control during chlorination. Proper disinfection requires sufficient mechanical application and volume addition to drive the chlorine chemistry into areas more often inhabited by anaerobic bacteria.

Chlorine dosage is important as excessive chlorine, even in pH controlled applications, causes oxidation of polysaccharide material (the exopolymer or slime produced by the bacteria). This oxidation or sealing off of the top layers of the biofilm results in some protection against the chlorine (see Figure 3.9) for the remaining bacteria.

In addition to selecting a chlorine dosage, the Michigan DEQ also asked for a method that would be readily available to the smaller contractor to control the pH during chlorination so that the more effective hypochlorous acid would be available for disinfection (Schnieders J., 1998). We tested hydroxyacetic acid (glycolic) among others and found it to work well for this application; however, this acid is not readily available. We then tested vinegar (acetic acid 5%). This worked well in the laboratory and one of the formulae in Table 5.1 was set up for small well application.

Table 5.1. Formulae for pH control of chlorination

$$\text{Number of gals. (3.8 liter) of vinegar} = \frac{\text{Alkalinity}}{100\text{ mg/l}} \times \frac{\text{Chlorine}}{500\text{ mg/l}} \times \frac{\text{Volume}}{100\text{ gal. (380 liter)}}$$

For larger systems:

$$\text{Number of qts. (liters of glycolic)} = \frac{\text{Alkalinity}}{100\text{ mg/l}} \times \frac{\text{Chlorine}}{200\text{ mg/l}} \times \frac{\text{Volume}}{1{,}000\text{ gal. (3800 liter)}}$$

CAUTION: Hypochlorite solution added to water with a pH below 5.0 will release chlorine gas, which is extremely toxic. This procedure should only be done in the open with good ventilation.

The workshops and discussion groups held around the State came up with some excellent information. Two of the important procedures, which appeared to have the most effect on clearing wells of coliforms, included good well development during the well construction phase and excess pumping of the well prior to chlorination. It was apparent time and time again that those well contractors who properly developed their wells had very few coliform problems. Also, some of these same contractors as well as numerous contractors and well owners (particularly larger wells) from across the state have reported amazing success by pumping the well for two to four weeks before chlorination. A recent survey in Michigan showed that of the wells which showed positive coliforms after initial chlorination, 60% were found coliform negative after pumping for 24 to 48 hours (Holben, 2001).

Minimum Recommendations for Well Chlorination

With the results of our workshop surveys and observations from our laboratory work, the following recommendations were formulated to improve well chlorination:

1) With new well construction or rehabilitation, sufficient time should be spent on well development.

2) Pump the well for an extended period of time before disinfection (minimum of 20 to 40 well volumes).

3) Prepare a chlorine solution (between 50 and 200 mg/l) above ground in sufficient volume to equal three to four times the standing well volume. The chlorine solution should be prepared by first adjusting the pH with a suitable acid or product such as ChloraPal by Design Water Technologies or NuWell-310 or NuWell-410 by Johnson Screens. Once the pH of the water is at 4.5 to 5.0 add the sodium hypochlorite and mix.

WARNING: This must be done in a well-ventilated area out in the open as chlorine gas can be produced. Once the sodium hypochlorite is added the pH will immediately rise and stop further chlorine release. (Chlorine is not released above a pH of 5.0.)

4) Place the prepared chlorine solution into the well and surge or swab into the gravel pack and immediate zones around the well. The surging time is dependent on how deep the well is, but try to surge for one to two minutes for every foot of well screen or open borehole below the water level. Jetting of the prepared chlorine solution may be preferred in open borehole wells. Time allotted should also provide one to two minutes per foot. Some surging and/or jetting should be done above the screen to make sure that casing surfaces are disinfected. Let the solution stand overnight.

5) The second day surge or jet for one-half the original time and evacuate the well starting just below the static water level. The more vigorous and continuous the evacuation the better. Evacuate until no chlorine is detected and total volume discharged equals a minimum of 20 well volumes. Evacuation must be continued to the well bottom in order to remove all debris that has accumulated.

To summarize what has been accomplished:

1) The development and extended pumping has restored not only the formation but has also removed any loose biofilm and other debris and returned other biofilm to a less active or more quiescent state. This means less free-swimming bacteria and less chance for a coliform positive test.

2) By injecting three to four well volumes of chlorine solution we can effectively flood the well environment; not only the casing but also the entire well bore and surrounding formation. This brings the chlorine solution in contact with the deeper zones where anaerobic activity would be present.

3) By limiting the chlorine dosage between 50 and 200 mg/l we are decreasing the possible oxidative reaction against the biofilm, thereby reducing the protection of deep-seated coliforms.

4) By using pH control we are converting all the hypochlorite ions to hypochlorous acid which is at least 100 times more biocidal than chlorinating with a hypochlorite solution.

5) By using sodium hypochlorite we are insuring complete solubility of the chlorine product even in hard water.

6) By surging the chlorine solution into the aquifer and letting it stand overnight we are insuring contact in all zones of the well with sufficient contact time for disinfection.

7) By thorough evacuation of pump out we are insuring the complete removal of chlorine and, to some extent more importantly, the biological or organic debris which would be a food source for future bacterial growth.

We have recommended these steps for many wells across the country and have had considerable success. It goes without question that the well has been completely investigated to ascertain that no contaminating source is supplying coliforms to the well. In our laboratory we have made a rule of enumerating the coliforms in order to ascertain the severity of the presence of the coliforms. While we haven't completed enough surveys to give a final value, currently we are using a count of one to two bacteria per milliliter to signal a possible contaminating source which must be found before disinfection can be successful. Counts less than 100 per 100 ml are believed to be environmental bacteria thought of as regularly being housed in the well and not being added to by outside contamination.

Pathogenic Bacteria in Wells

If we use the information learned about coliforms in the well system, we should be able to picture the residency of a pathogen in a well structure. Using this information, a more realistic method can be engineered to remove a pathogen or other microbial contaminant from a well through more directed cleaning and disinfection. In Chapter 2 a list of potential pathogenic bacteria found in waste material that may enter our water systems is provided together with the usual environmental requirement for growth (Holt and others, 1994). More than 90% of the organisms listed are capable of living in an anaerobic environment. While it would be expected that many of these organisms would be far more susceptible to environmental stress, many similarities between them and coliforms exist as to their potential methods of survival or inhabitation in a well environment. Thus we would expect these particular bacteria to enter the anaerobic dominant biofilms deeper in the anoxic areas of the well structure. Work done by Marsh and Bradshaw and reported in 1996 showed that fastidious organisms are capable of forming small metabolically structured units within the biofilm that allow them to survive and proliferate under otherwise stressful conditions.

Other Pathogens in Wells

Along with the list of pathogenic bacteria that could enter our water supplies, Rose and Grimes (2000) also included in their appendix listings other microbial pathogens that could enter our water (see Appendix A). Of these, the human viruses, pathogenic fungi, cyanobacteria and protozoa are potential well inhabitants. While most of these organisms are fastidious and would be stressed by the environment, little has been reported on how they reside in the environment. Protozoans often feed on biofilm and the bacteria within, while viruses are in search of host cells, and fungi would also look to the metabolically rich biofilm structures as a good feeding ground.

Many of these named organisms would probably not proliferate in the well environment, however they all could enter the biofilm and either temporarily or permanently reside there. From this location they would be able to periodically enter the water flow resulting in the direct contamination of our water, or more directly enter the user and may result in a disease process depending on the organisms and the susceptibility of the host.

Removal of Pathogens

In the above section some evidence was given to the most probable residing places for pathogens in a well environment. In Chapter 2 a discussion of the dynamics of a well was provided. It was reasoned that microorganisms, in order to stay in a well environment, would have to find a hiding place or residing structure, otherwise they would be washed from the well system. This comparison to the coliform bacteria, which are capable of eluding numerous chlorine applications because of their ability to enter deep-seated anaerobic dominant biofilms, is the key to developing a good decontamination or disinfection program. The same steps that are used to successfully remove the coliforms should be incorporated in a strong decontamination procedure. Simple chlorination of wells with minimal mechanical activity has been only partially successful in removing coliforms; therefore, the same procedure or similar approaches will fail with decontamination attempts regardless of the strength of the disinfecting chemistry (Schnieders J, 2001). A quick review of the procedures that must be included in the decontamination process follows if serious contamination with microorganisms exists.

Recommended Decontamination Procedure

1) The pump must be removed.

2) If water analysis, age of the well, and experience with aquifer water and well operation reveal the probability of heavy mineral and biological fouling, then the well should be cleaned initially with an acid and a biodispersant. (See Chapters 9, 10, and 11 for a review of this process). This procedure is designed to limit the areas of hiding and attachment in the flow paths of the well.

3) The well should be thoroughly evacuated with removal of all debris and development of the well formation. Particular attention should be paid to sumps or dead legs, if present, as these often harbor heavy anaerobic growth (Schnieders M, 2001). The well development carried out by continual evacuation over the entire area of the screened section in gravel packed wells and over the open area in open borehole wells will remove much of the loose biofilm and other debris from the gravel pack and/or formation if the acid cleaning is not performed.

4) A sodium hypochlorite solution should be prepared in a tank in a volume equal to four times the standing well volume. The hypochlorite concentration should approximate 200 mg/l and the pH should be adjusted to a pH of 6.5 (see Chapter 9 and Table 5.1).

5) The prepared chlorine solution should be tremied into the well starting at the bottom and working upward. If the well is an open borehole construction, then a jetting tool may be used to apply the chemistry and at the same time jet the surface of the casing and borehole wall. The jetting rate and pressure should be adjusted to spend sufficient time at each interval to clean the surface and to penetrate the openings leading to the fractures feeding the well. In some cases the solution can be circulated back to the surface and after removal of solids again jetted into the well to extend the application procedure.

6) In wells which are gravel packed, after the chlorine solution has been applied, a double surge block or in smaller wells a tight fitting swab should be used to surge the solution into place. Spend at least two minutes per foot of screen in order to clean the screen and casing surface and penetrate the gravel pack.

7) Once the surging has been completed test the solution to track the use of the chlorine. If the chlorine level has been depleted below 100 mg/l then additional pH controlled chlorine solution should be added. The system should be allowed to set idle overnight.

8) The casing above the water level, the pump and any down piping, as well as adjacent structure should be thoroughly cleaned and washed with pH adjusted 200 mg/l hypochlorite solution.

9) The next day resurge or jet the well for a time period equal to one-half the previous days effort and begin to evacuate the well. Start just below the static water level and moving downward continually evacuate each level until clear. Evacuation must reach the well bottom to insure all debris is removed.

10) After pumping from the well, neutralize the chlorine cleaning solution before discharge.

CHAPTER 6 Mineralization Potential

Every well has a potential for mineral accumulation within its environment dependent on the chemistry and biology of the aquifer water, the structure of the well, and the operating parameters. The presence, and to a greater extent the formation of these minerals, may play a significant role in the contamination process and the decontamination of the well. While there are a multitude of minerals which could form in any given well, there are only a few of sufficient probability capable of producing enough mass that their presence will interfere with well cleaning. These are the carbonates, the sulfates, and the oxides. Occasionally phosphates are formed in a well, usually the result of over treatment or more often improper treatment with this type of chemistry. Sulfides are also a minor formation often seen in wells, however, occasionally they can become a dominant deposit. The mineralization of the flow pathways or surfaces in a well often follows the fouling of these areas with biofilm formation as reviewed in Chapter 3.

Chemical Reactions in the Well Environment

Each of the potential types of mineral formation has specific chemistries that influence their deposition. These chemistries are usually initiated within the well environment due to certain physical changes (such as pumping rate increases that influence the chemical profile) or certain biochemistry changes which are initiated usually by bacterial growth within the environment. Other processes, such as the availability of iron due to corrosion of the well structural components, will be discussed.

There are two deposits which dominate mineral accumulation in wells and I would not do justice if I picked one over the other. Both are significant in that they produce severe blockage in conjunction with biofouling and both occur very often throughout the country, though they are usually more specific to the respective aquifer water chemistry. One is calcium carbonate formation, which is seen most often in the areas of high groundwater hardness and consequently high alkalinity, and the other is the build up of iron oxides which usually occurs in areas of low groundwater pH and iron bearing water. Steel casing and screen corrosion may add to the accumulation.

Saturation Index

Langelier, in his 1936 work, developed the Saturation Index (SI) that is a measure of a solutions' ability to dissolve or deposit calcium carbonate. In developing the SI, Langelier derived an equation for the pH at which water is saturated with calcium carbonate (pH_S). This equation is based on the equilibrium expressions for calcium carbonate solubility and bicarbonate dissociation. To approximate actual conditions more closely, pH_S calculations were modified to include the effects of temperature and ionic strength. The Saturation Index is defined as the difference between actual pH and calculated pH_S. The magnitude and sign of the SI value shows water's tendency to form or dissolve a deposit. Because water that is depositing carbonate scale (high alkalinity, neutral pH or higher) is usually not corrosive to steel, and water that will dissolve carbonate deposits may be corrosive, the index can also be used as a general indicator of the corrosivity of the particular water.

Although information obtained from the SI is not quantitative, it can be useful in estimating water tendencies toward calcium carbonate deposition and its corrosive or dissolution properties toward mineral deposits or well components. Since the deposition of sulfates occurs under most of the same parameters, a positive index, with sulfates in excess of at least 100 mg/l, is taken as an indication of possible calcium sulfate deposition.

The index can be calculated using the simple formula:

$$pH_{actual} - pH_s = SI \qquad \text{where} \qquad pH_s = A + B - C - D$$

The values for A, B, C, & D are then obtained from a table where A is a value given for the total dissolved solids concentration, B is a value representing the temperature, C is the calcium concentration expressed as mg/l $CaCO_3$, and D is the total alkalinity expressed as mg/l $CaCO_3$ (see Table 6.1 through 6.3).

Table 6.1. Saturation Index Total Solids/Temperature

A	
Total Solids	
mg/l	
50	0.07
75	0.08
100	0.10
150	0.11
200	0.13
300	0.14
400	0.16
600	0.18
800	0.19
1000	0.20
2000	0.23
3000	0.25
4000	0.26
5000	0.26
6000	0.28

B					
Temperatures in °F					
	Units				
Tens	0	2	4	6	8
30		2.60	2.57	2.54	2.51
40	2.48	2.45	2.43	2.40	2.37
50	2.34	2.31	2.28	2.25	2.22
60	2.20	2.17	2.14	2.11	2.09
70	2.06	2.04	2.03	2.00	1.97
80	1.95	1.92	1.90	1.88	1.86
90	1.84	1.82	1.80	1.78	1.76
100	1.74	1.72	1.71	1.69	1.67
110	1.65	1.64	1.62	1.60	1.58
120	1.57	1.55	1.53	1.51	1.50
130	1.48	1.46	1.44	1.43	1.41
140	1.40	1.38	1.37	1.35	1.34
150	1.32	1.31	1.29	1.28	1.27
160	1.26	1.24	1.23	1.22	1.21
170	1.19	1.18	1.17	1.16	

Table 6.2. Saturation Index Calcium Hardness

C

Calcium Hardness Expressed as mg/l $CaCO_3$

(for 3 to 209 mg/l $CaCO_3$ use upper table; for 210 to 990 mg/l $CaCO_3$, use lower table)

	Units									
10s	**0**	**1**	**2**	**3**	**4**	**5**	**6**	**7**	**8**	**9**
0				0.08	0.20	0.30	0.38	0.45	0.51	0.56
10	0.60	0.64	0.68	0.72	0.75	0.78	0.81	0.83	0.86	0.88
20	0.90	0.92	0.94	0.96	0.98	1.00	1.02	1.03	1.05	1.06
30	1.08	1.09	1.11	1.12	1.13	1.15	1.16	1.17	1.18	1.19
40	1.20	1.21	1.23	1.24	1.25	1.26	1.26	1.27	1.28	1.29
50	1.30	1.31	1.32	1.33	1.34	1.34	1.35	1.36	1.37	1.37
60	1.38	1.39	1.39	1.40	1.41	1.42	1.42	1.43	1.43	1.44
70	1.45	1.45	1.46	1.47	1.47	1.48	1.48	1.49	1.49	1.50
80	1.51	1.51	1.52	1.52	1.53	1.53	1.54	1.54	1.55	1.55
90	1.56	1.56	1.57	1.57	1.58	1.58	1.58	1.59	1.59	1.60
100	1.60	1.61	1.61	1.61	1.62	1.62	1.63	1.63	1.64	1.64
110	1.64	1.65	1.65	1.66	1.66	1.66	1.67	1.67	1.67	1.68
120	1.68	1.68	1.69	1.69	1.70	1.70	1.70	1.71	1.71	1.71
130	1.72	1.72	1.72	1.73	1.73	1.73	1.74	1.74	1.74	1.75
140	1.75	1.75	1.75	1.76	1.76	1.76	1.77	1.77	1.77	1.78
150	1.78	1.78	1.78	1.79	1.79	1.79	1.80	1.80	1.80	1.80
160	1.81	1.81	1.81	1.81	1.82	1.82	1.82	1.82	1.83	1.83
170	1.83	1.84	1.84	1.84	1.84	1.85	1.85	1.85	1.85	1.85
180	1.86	1.86	1.86	1.86	1.87	1.87	1.87	1.87	1.88	1.88
190	1.88	1.88	1.89	1.89	1.89	1.89	1.89	1.90	1.90	1.90
200	1.90	1.91	1.91	1.91	1.91	1.91	1.92	1.92	1.92	1.92
100s	**0**	**10**	**20**	**30**	**40**	**50**	**60**	**70**	**80**	**90**
200		1.92	1.94	1.96	1.98	2.00	2.02	2.03	2.05	2.06
300	2.08	2.09	2.11	2.12	2.13	2.15	2.16	2.17	2.18	2.19
400	2.20	2.21	2.23	2.24	2.25	2.26	2.26	2.27	2.28	2.29
500	2.30	2.31	2.32	2.33	2.34	2.34	2.35	2.36	2.37	2.37
600	2.38	2.39	2.39	2.40	2.41	2.42	2.42	2.43	2.43	2.44
700	2.45	2.45	2.46	2.47	2.47	2.48	2.48	2.49	2.49	2.50
800	2.51	2.51	2.52	2.52	2.53	2.53	2.54	2.54	2.55	2.55
900	2.56	2.56	2.57	2.57	2.58	2.58	2.58	2.59	2.59	2.60

Table 6.3. Saturation Index Alkalinity

D

Alkalinity Expressed as mg/l $CaCO_3$

(for 1 to 209 mg/l $CaCO_3$ use upper table; for 210 to 990 mg/l$CaCO_3$, use lower table)

10s	0	1	2	3	4	5	6	7	8	9
0		0.00	0.30	0.48	0.60	0.70	0.78	0.85	0.90	0.95
10	1.00	1.04	1.08	1.11	1.15	1.18	1.20	1.23	1.26	1.29
20	1.30	1.32	1.34	1.36	1.38	1.40	1.42	1.43	1.45	1.46
30	1.48	1.49	1.51	1.52	1.53	1.54	1.56	1.57	1.58	1.59
40	1.60	1.61	1.62	1.63	1.64	1.65	1.66	1.67	1.68	1.69
50	1.70	1.71	1.72	1.72	1.73	1.74	1.75	1.76	1.76	1.77
60	1.78	1.79	1.79	1.80	1.81	1.81	1.82	1.83	1.83	1.84
70	1.85	1.85	1.86	1.86	1.87	1.88	1.88	1.89	1.89	1.90
80	1.90	1.91	1.91	1.92	1.92	1.93	1.93	1.94	1.94	1.95
90	1.95	1.96	1.96	1.97	1.97	1.98	1.98	1.99	1.99	2.00
100	2.00	2.00	2.01	2.01	2.02	2.02	2.03	2.03	2.03	2.04
110	2.04	2.05	2.05	2.05	2.06	2.06	2.06	2.07	2.07	2.08
120	2.08	2.08	2.09	2.09	2.09	2.10	2.10	2.10	2.11	2.11
130	2.11	2.12	2.12	2.12	2.13	2.13	2.13	2.14	2.14	2.14
140	2.15	2.15	2.15	2.16	2.16	2.16	2.16	2.17	2.17	2.17
150	2.18	2.18	2.18	2.18	2.19	2.19	2.19	2.20	2.20	2.20
160	2.20	2.21	2.21	2.21	2.21	2.22	2.22	2.23	2.23	2.23
170	2.23	2.23	2.23	2.24	2.24	2.24	2.24	2.25	2.25	2.25
180	2.26	2.26	2.26	2.26	2.26	2.27	2.27	2.27	2.27	2.28
190	2.28	2.28	2.28	2.29	2.29	2.29	2.29	2.29	2.30	2.30
200	2.30	2.30	2.30	2.31	2.31	2.31	2.31	2.32	2.32	2.32

100s	0	10	20	30	40	50	60	70	80	90
200		2.32	2.34	2.36	2.38	2.40	2.42	2.43	2.45	2.46
300	2.48	2.49	2.51	2.52	2.53	2.54	2.56	2.57	2.58	2.59
400	2.60	2.61	2.62	2.63	2.64	2.65	2.66	2.67	2.68	2.69
500	2.70	2.71	2.72	2.72	2.73	2.74	2.75	2.76	2.76	2.77
600	2.78	2.79	2.79	2.80	2.81	2.81	2.82	2.83	2.83	2.84
700	2.85	2.85	2.86	2.86	2.87	2.88	2.88	2.89	2.89	2.90
800	2.90	2.91	2.91	2.92	2.92	2.93	2.93	2.94	2.94	2.95
900	2.95	2.96	2.96	2.97	2.97	2.98	2.98	2.99	2.99	3.00

Calcium Carbonate Formation

Calcium carbonate, or calcite, is formed from the calcium portion of the hardness level in the water and the bicarbonate ion (alkalinity) present. Most often calcium carbonate (calcite) is the dominant carbonate form but occasionally magnesium carbonate is formed in a dolomite environment. Such an environment is one in which the two cations, calcium and magnesium are present in or near the same concentration. Usually magnesium carbonate is not found unless this condition exists. Where iron levels are high, iron carbonate is often included in the deposit. As was reviewed earlier, certain conditions need to exist for carbonate formation to take place. All these conditions in some way are covered in the Saturation Index. If any of the parameters of the Index are raised then precipitation of a carbonate usually takes place. The following are the key parameters of the Index:

- Alkalinity concentration
- Calcium concentration
- Total Dissolved Solids concentration
- pH
- Temperature

Key to Determination:

Run Saturation Index: A positive number indicates you could have some deposits formed in the well. (A negative number may indicate the potential for some corrosion activity.) Deposits of carbonate scale can be the primary blockage if pH is >7.5, hardness is >250 mg/l or calcium above 175 mg/l, and alkalinity is >220 mg/l.

Calcium Sulfate Formation

As with carbonate formation calcium sulfate, or gypsum, also depends on the calcium side of the hardness reading. Gypsum formation depends on the presence of sulfate in the water usually at the 100 mg/l level if sufficient calcium and alkalinity are present. Testing values to check for in addition to those for carbonate potential are:

- Carbonate Hardness
- Non Carbonate Hardness
- Sulfates

Key to Determination:

If the Saturation Index is positive indicating the potential for calcite deposits, the non carbonate hardness is greater than the carbonate hardness, and the sulfate levels are in excess of 100 mg/l then gypsum or sulfate scale could deposit on the screens and in the gravel pack or formation (especially if there is an increase in pumping rates or aquifer pH increase). Sulfate scale may be the major blockage if pH is >7.5, calcium >175 mg/l, alkalinity is <1/2 of calcium concentration, and sulfates are >150 mg/l.

Oxides Formation

Oxides are most often an oxidized form of the metals iron and manganese. Because iron oxides are often the form of corrosion by-products, the potential may require not only knowledge of the existence of iron and manganese levels but also the potential for corrosion in the well. Therefore, run all the above parameters for carbonate and sulfate potential plus:

- Iron
- Manganese
- ORP (Oxidation Reduction Potential)

Key to Determination:

If the Saturation Index is negative and/or the ORP is below 150 mV, then you have corrosive water and any metal in the system is subject to releasing ions (most likely iron) into the water to become oxidized to a metal oxide. The red and black accumulations in your well are a product of this reaction. You may also accumulate iron and manganese oxides, which are also red (usually iron) and black (usually manganese), from bacterial oxidation of the iron and manganese found in the aquifer water either there naturally or as a result of corrosion activity. Serious fouling can occur when iron levels are in excess of 1.0 mg/l and manganese is above 0.1 mg/l. Brucite or magnesium hydroxide may also form in the type of environment conducive to calcite deposits if the magnesium calcium ratio is >1:1, the alkalinity is >220 mg/l, and the total hardness is >250 mg/l.

Physical and Biological Changes

In addition to the primary chemical parameters controlling the respective deposits listed above, there are many secondary phenomena that control the deposition of these minerals and a few others of importance. Here are a couple of examples.

Degassing or Carbon Dioxide Release

Wells that have operated for prolonged periods often develop severe blockage and loss of capacity suddenly or over a relatively short period of time. We see this particularly when pump size and/or pumping capacity has been increased in order to supply more water during periods of high water demand.

The concentration of carbon dioxide in the aquifer water varies directly with the amount of pressure in the subsurface. Since calcium and iron bicarbonates are carried in the groundwater in proportion to the amount of dissolved carbon dioxide present, an increase in pressure would permit more bicarbonate to be present. Thus an increased level of carbon dioxide would permit more calcium and iron in solution.

When water is pumped from the aquifer however, the opposite reaction takes place. First, the piezometric head declines lowering the hydrostatic pressure on the groundwater which results in a dramatic change in the chemical equilibrium at the well screen/aquifer interface. Due to the pressure drop carbon dioxide is released or “degassed” at or near the screens and the soluble bicarbonates in the groundwater convert to the insoluble carbonate form. The carbonates are then deposited in the pore spaces and on the screen plugging the flow areas and reducing the capacity yield of the well (Lehr and others, 1988). The chemical reaction is as follows:

$$Ca(HCO_3)_2 \rightarrow CaCO_3\downarrow + CO_2\uparrow + H_2O$$

$$Fe(HCO_3)_2 \rightarrow FeCO_3\downarrow + CO_2\uparrow + H_2O$$

Hydrogen Sulfide Production

In wells supplied by aquifers high in sulfates we often see the development of large zones of sulfate-reducing bacteria. These bacteria are usually located in the lower areas of the well and are some of the most prolific of the anaerobic organisms. The reduction chemistry involved is a series of complex biochemistry reactions resulting in the corresponding production of the sulfur ion (S^{-2}) and the hydrogen ion (H^+) resulting in the formation of hydrogen sulfide gas (H_2S) which then evolves from the site moving upward through the well (Chapelle, 1993). While the reduction chemistry ultimately produces acid (low pH conditions corrosive to the well structure), hydrogen sulfide gas is the most noticed due to its offensive odor (rotten egg odor).

CAUTION: Saturated water solutions of the gas can produce an acid pH of 4.5 and air gas mixtures as low as 4.3% are explosive. The gas is heavier than air so wells located in low areas or vaults should be well marked if the gas is present.

In addition to sulfide formation, the sulfate-reducing bacteria are often the cause of large quantities of iron oxide, typically ferric oxide (Fe_2O_3), a reddish brown deposit, as well as ferric ferrous oxide (Fe_3O_4), a semi-magnetic brown-black deposit. These reactions follow the corrosion effect on the casing and screen and the resulting oxidation of the iron by the oxygen in the water, particularly higher up in the water column.

Another phenomenon that should be again mentioned at this point since it directly follows the formation of sulfides by the sulfate-reducing bacteria, is the proliferation of sulfur oxidizing bacteria, such as the *Thiothrix*. This bacteria oxidize sulfides to sulfates and in the case of hard water can provide a mechanism for gypsum (calcium sulfate) formation. Since *Thiothrix* are usually found in flowing sulfide-containing water, this organism is known to cause floating black particles, actually rosettes (see Figure 4.4). The rosettes are formed by the growth of the organism in the distribution water. The resulting bio-accumulation is the primary cause of sulfate formation as part of the deposits seen in these systems.

While *Thiothrix* are one of the more distinguishing bacteria that oxidize sulfides, there are other sulfur oxidizing bacteria such as the *Thiobacillus* which are capable of oxidizing a wide range of reduced sulfur compounds of which sulfates are the end product. These bacteria are also wide spread in soil and fresh water environments as well as distribution piping in water and waste water systems and are responsible for major acid corrosion problems in these systems.

CHAPTER 7 Geological Considerations

The chemistry and microbiology of the aquifer are not the only factors which control blockage of a well system. The geology present around a well can be a controlling mechanism for the initial formation of incrustation or the matrix of both mineral deposits and bacterial biofilm The geology not only determines the well design but also forms the base foundation for both or either of the blockage forming mechanisms. The aquifer formation, to some extent, may be the controlling factor in the placement and type of blockage and consequently must be taken into consideration when a well system is being cleaned.

Well systems are engineered to draw water from an aquifer and since there are many different types of aquifers there are many variations in well design. In general, though, they are constructed in either consolidated or unconsolidated geological formations which determines the main engineering design. Those wells constructed in continuous consolidated lithology are often open borehole construction. That is the walls of the drilled hole in consolidated material are firm enough to be self-supporting and no casing or screen is required to prevent the collapse of the well structure. These types of wells usually have a casing in the upper part to house the pump and to facilitate the installation of grout or a proper seal to prevent surface water from infiltrating the well.

The second general type of well construction is the gravel pack well which is the well of choice in unconsolidated aquifers (Figure 7.3). This type of well uses a natural or artificial gravel pack, casing, and a screen, slotted, or perforated pipe to support the borehole wall and provide some additional filtration of the incoming water. These are the common large wells of the southwest and alluvial aquifers throughout the country.

The Consolidated Aquifer

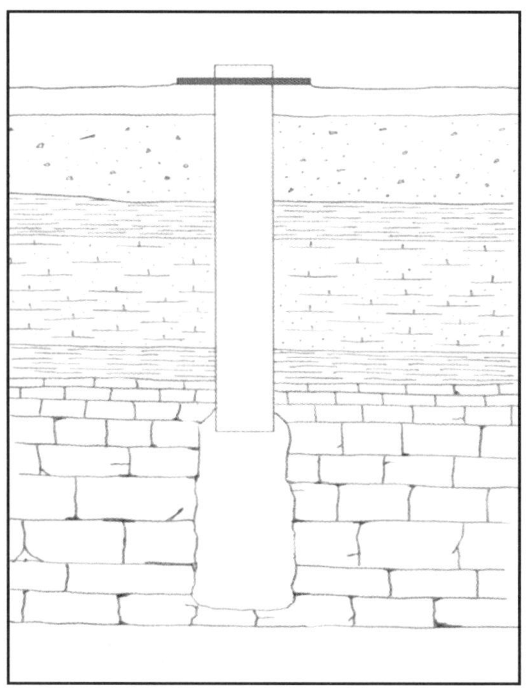

Figure 7.1. Open borehole well in consolidated aquifer

As opposed to the multitude of openings of the porous sand and gravel formations, consolidated material is relatively non-porous and water flow into the well depends on the flow through joints and fractures in the solid formation. These joints or fractures are subject to plugging both from the debris generated during construction or from mineral precipitation and biological buildup as the well is operated.

While the cleaning chemistry for the open borehole wells does not vary much from the cleaning of traditional screened and/or gravel pack wells, the mechanical methods are somewhat different. Due to the possibility of damage to the borehole wall, surge blocks and swabs are not used, at least not of the tight fitting variety. A good method of both cleaning and application of the chemical in this type of well is the use of jetting tools which allow a controlled application sufficient to wash the borehole surface and to clean the crevices or openings. Since these openings or flow pathways are not straight or exactly perpendicular to the well bore, the jetting stream is used primarily to open the outlet of the fracture or water passageway. Once open, the cleaning chemistry is allowed to flow to the deeper areas. The subsequent pump-out or airlift extraction actually develops the deeper cleaning. The alternating application of these two technologies, as well as the use of tools designed to simultaneously accomplish both operations, is considered one of the better methods of both well cleaning and well development for the open borehole constructed well. During initial construction well development is used to remove the debris and to attempt to enlarge the fractures if possible. In this way the contractor is either trying to return the aquifer to as close to the original condition or free flowing characteristic or enhance the flow of the aquifer water to the well.

There are other ways to enlarge and deepen the openings such as the use of explosives or hydrofracing, where water is pumped into the well under high pressure. To restore the aquifer and to apply the cleaning chemistry, however, it is still jetting and evacuation that is used most often.

In open borehole wells where uneven distribution of fractures or flow pathways exist, particularly in lower sections of the well, flooding the well with chemistry or even the jetting of the well may not provide adequate cleaning. In these situations a considerable volume of the cleaner can be lost through the more open fractures leaving little to be recirculated or even penetrate the plugged openings (see Figure 7.2). A larger volume of cleaning solutions (four to five times the borehole volume) of a more dilute concentration should be jetted over an extended period.

This will not only provide more even cleaning of the borehole face but also a more equal penetration of the aquifer formation throughout the exposed borehole area.

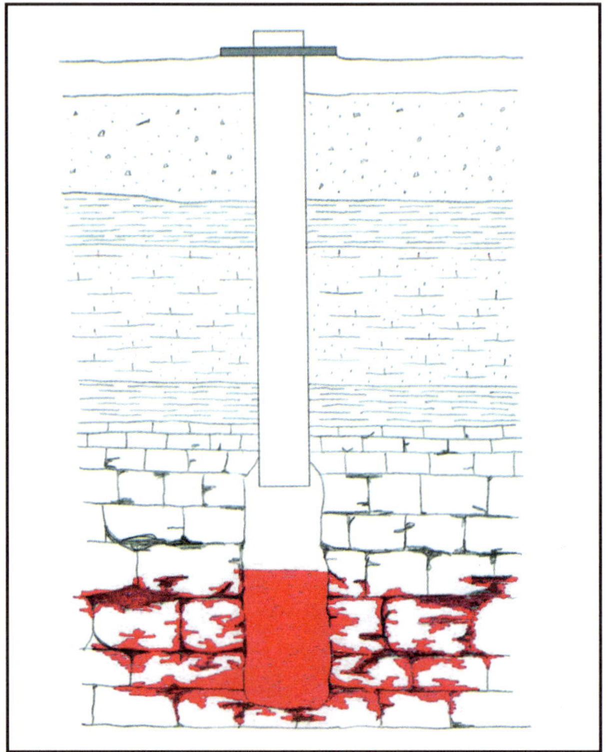

Figure 7.2. Open borehole showing loss of cleaning chemistry Often large volumes are lost to chambers adjacent to well.

The Unconsolidated Aquifer

Installing a well into an unconsolidated, alluvial formation requires a different type of construction. Here support for the borehole wall must be provided both during the construction phase and later as a form of permanent structure. The gravel pack and screen, slotted or perforated pipe are used as the permanent well structure or support for the borehole wall following drilling completion.

The gravel pack well may be constructed with the use of drilling mud, synthetic or natural, that is used to support the borehole wall during construction. Loose soils and gravels would quickly fall inward if a mud layer was not applied during the drilling process. Once the drill bit, etc. is removed, casing and screen (sometimes in alternate layers) is centered in the hole and a uniform gravel is then put into the space between the mud caked borehole wall and the screen, slotted or perforated pipe. Usually cement is needed to seal the space above the gravel pack and screen area and in between the casing and the borehole wall. In some cases cement is used in blank areas between aquifers so that a channel (via the gravel pack) is not opened between two separate aquifers. Well development with the use of surge blocks or swabs or jetting is then carried out to clean and remove the mud and cuttings or drilling debris from the borehole wall thereby opening the flow pathways for water.

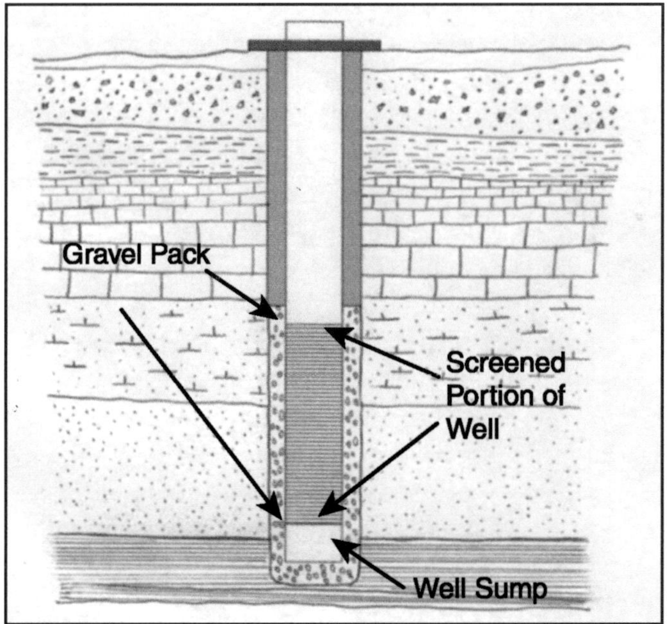

Figure 7.3. Gravel pack well in unconsolidated aquifer

Well development is also required with wells constructed using air rotary drilling because, while it is not necessary to remove mud, considerable debris in the form of cuttings must be removed to prevent flow blockage. The developmental process also rearranges the formation returning it to a more natural setting for the gravels, sands, etc. that make up the aquifer. The development is enhanced with the use of polymeric mud control chemicals or phosphates. Phosphates have been used effectively for many years but often adversely affect the formation. This occurs by swelling clay layers and in certain waters leaving a phosphate precipitate which can stimulate bacterial growth. The polymeric chemistries that have come on the market are often phosphate free and provide a much-needed improvement in mud removal.

A research project at our laboratory designed to study the characteristics of several of the new mud removal polymers uncovered some interesting and very useful information concerning mud removal. We had been using a particular brand of bentonite to mud the borehole wall of our laboratory wells to test the efficiency of the mud removal polymers. Most companies routinely recommend chlorinating wells after synthetic (those made from polyacrylamide) muds have been used in order to break the synthetic polymer before development is started. Manufacturers also advertise that some of the bentonite muds are "fortified" with these same polyacrylamide polymers. Since we were not having much success at mud removal, we decided to try a chlorine application prior to development even though the product we were using was not suppose to contain a fortifier. We used varying levels of hypochlorite for two to six hour periods but still did not successfully break the mud application. Then, almost by accident, we left several wells set overnight and noticed a much improved effect on the mud. After a number of trials we discovered that it takes 1200 to 1500 mg/l chlorine eight to 12 hours to fully break down the bentonite application. After some additional testing we discovered that almost all the bentonite muds on the market, even those that do not advertise any additives, appear to have at least 0.02% acrylamide polymer present. This amount appears to be sufficient to require a higher chlorine wash. Since that time we have recommended an overnight chlorination at 1500 mg/l followed by good development with one of the mud active polymers like Mud Buster by Design Water Technologies or NuWell-220 manufactured by Johnson Screens. Using the increased chlorine levels and the extended application period, many contractors have been able to remove muds from wells over 15 years old.

Incidentally, the two products mentioned tested five to ten times more efficient at mud removal than all five of the phosphate products tested.

Well Development

Well development is a very important part of well construction. It is designed to remove the mud from the borehole wall drawing it through the gravel into the well for removal from the well system. In open borehole wells it removes cuttings and opens and deepens fractures and fissures to insure better water flow to the well. Without development the areas of flow, the fractures and pore spaces would not be open or at least could be substantially blocked to water flow into the well. Just as the development method for open borehole wells is useful during the cleaning of crevasse openings, the methods of development adaptable for gravel pack wells are often the best for the chemical cleaning process of those wells. The surging and evacuation of water from the well is able to clean the borehole wall indicating that this process with proper chemistry can be used successfully for rehabilitation and cleaning.

During our research work in Michigan we had the opportunity to talk with many well drillers concerning well development and its relationship to a good producing well. At the time we were discussing the occurrences of coliform positive wells, especially new wells. One contractor stood out in the group because he had never had a positive coliform test following new well construction. I invited him to tell us his secret. He hesitated a little and said he would tell us how he built the well as he didn't know he possessed a secret. He started with, *"My grandfather always told me that it takes twice as long to develop a well as it does to drill it."* In that sentence he explained his success. Clearing the aquifer of debris is first of importance in removing any offending microorganism from a well. He went on to describe a very good method for small well development driving home the importance of good well development. As many drillers present had become complacent in their well development due to the abundance of the local aquifers, it became apparent that good well development has other advantages than just increasing yield.

Redevelopment

Redevelopment is necessary in wells due to the presence of blockage or fines from formation or geological changes rather than those due to biological and mineral deposits.

Sand, clays, and particulate matter from the formation are usually considered in this category. Other substances that are not geological but might fall into this group as physical blockage are bentonite and some of the synthetic polymers used as mud during the well construction period. As water moves toward the well, sand and fines are moved within the higher velocity developed in the well. Eventually these fines and sands impact the gravel pack and even enter the casing area. Sand, of course, can hurt the pump resulting in erosion corrosion of pump parts and even screen and slot openings. Larger openings lead to more sand feeding into the well.

Clays also move toward the well especially if they make up a large part of the formation and/or water velocity is increased by increased pumping rates. Clays as well as bentonite and synthetic mud left in the well during construction all can become part of blockage especially in the gravel pack area.

Sand traps are a way of measuring the sand producing rate of a well and an increasing production should alert the operator to damage to screens and/or the possible infiltration of the gravel pack and the eventual loss of water production. Microscopic examination of water samples drawn from a pumping well will often give evidence of bentonite and natural clays accumulating in the well. Of course proper mud placement and well development during well construction will go a long way to keeping a clean gravel pack, however a repeat of the development may become necessary as the well ages and physical blockage begins to build up in the gravel pack and immediate formation. The redevelopment is necessary to return the well to its original formation structure and open flow areas.

The Differences of Aquifers

Bacterial growth is influenced in the downhole environment by many phenomena; however two of the more important hydrogeological features that lead to increased microbial growth are the "nature of the geological stratum" and hydraulic conductivity (Mansuy, 1999). Hydraulic conductivity accounts for the movement of water in the aquifer and bacterial growth in high transmissive zones has been reported to be much higher than in zones of low groundwater flow. The increased flow to an area brings with it increased nutrient loading and the removal of waste by-products, both of which are necessary to sustain a high bacterial population.

The nature of the geological stratum greatly encourages or discourages the development of bacterial growth. In alluvial or sand and gravel aquifers bacteria have numerous surfaces on which to establish biofilms. Water flow is greater and available food much more accessible. Such aquifers tend to produce very clean water and are capable of removing contaminants before they reach the well environment. Removal is accomplished simply by providing surface area for bacteria to reside or set up housekeeping. These biofilms tend to harbor a foreign organism until competitive bacteria starve them out or grazing protozoans consume them. This natural self-cleaning phenomenon of alluvial aquifers is very efficient, often removing a contaminant (both bacteriological and chemical) before it reaches a well use area.

In river alluvium the formation provides the same multiple surfaces for biofilm production as those in sand and gravel wells not under the influence of a river. In these aquifers, however, two phenomena control the biogrowth and account for the seasonal plugging of wells in these areas. Water flow from a very bacterially populated water source continues to flow toward the well in the season of abundant rainfall or recharge. During this period the biofilm becomes very populated and mature discharging considerable biomass into the flow. This discharge can lead to rapid fouling of the aquifer near or in the well environment. Later in the year the aquifer may be called on to feed the river and the directional change in the water hydraulics results in starvation of the biofilm which may also adversely affect the well production. Biofilm starvation also results in slough off and the biofilm formation on the well side opposite the river will severely impact that side of the well.

Consolidated formations on the other hand provide cracks and crevices for water movement. Surface area is greatly reduced and travel is much quicker since water is confined to a tight area through which to migrate. Biological and chemical contaminants that enter this type of aquifer have little time to interact with resident biofilms or to be absorbed onto surfaces. Well systems in this type of geological formation are more subject to contaminant infiltration. When these formations become fouled with bacterial growth they are difficult to clean. Fouling, however, is usually confined to the first few inches of the fracture or pathway. Here water slows down as it enters the well bore and is subjected to oxygen and nutrient availability in the standing water. Jetting application of cleaning and disinfectant chemistry is much more efficient since the chemical solution is forced into the opening cleaning the entrance way to the fracture and flow paths.

Glacial tills are another formation that can be somewhat detrimental to providing clean water. These formations have considerable clay content and often the clays are present in minute lenses offering excellent hiding zones for anaerobic growth. Wells in this type of formation should be routinely cleaned so that anaerobic activity is not allowed to build up. Once the growth patterns reach out from the well the area is extremely difficult to clean.

Karst formations, where erosion has provided large open areas and chambers near wells, can provide excellent opportunity for biological fouling. These areas can become fouled much like sinuses which continually drain contaminants back into the well. These systems are best protected by a good preventative maintenance program. The well should be maintained clean of debris so that organic deposits which encourage anaerobic growth are not allowed to build up in the adjacent chambers. Once heavily fouled, this type of formation is almost impossible to clean and any rehabilitation would require addressing the formation from a distance. A satellite well strategically placed may provide the access if the area of concern can be determined.

The examples given above are used to show the diversity of the aquifer involvement in well operation. They can also be used to show the diversity of wells and the need to treat each well individually. Of all the failures that our laboratory has tracked in the last 15 years most of them are due to the mistaken idea that all wells, or most in a given area, are alike. Each well is different and the blind treatment of wells will ensure failure or at the very least inefficient operation. Proper treatment and operation of a well requires that the well structure, as well as the chemical, biological, and geological nature, be known.

CHAPTER 8 Laboratory Testing

More often than not the testing of water samples from fouled wells, other than those tests for a particular contaminant, are easily forgotten and usually shrugged off as not being of much concern. On the contrary, laboratory testing of well water can be more important and more enlightening as to the well environment than all other investigative endeavors. In fact, if the renowned well videoing procedure had to be given up in place of a water analysis, take the laboratory tests. Chemical and biological testing of the well water can differentiate the types of chemical and biological activity which has resulted in well blockage. In the case of contaminants it can indicate the areas of accumulation, or in the case of pathogens the area of habitation thereby identifying areas that will require attention.

There are many analytical tests run on water, but usually we see testing utilized to satisfy public health tests for quality and to meet government requirements. Contaminant testing is used as well to satisfy water quality and the need to meet regulations. Unfortunately these test results are checked for compliance and very few are used to assess the condition of the well and the chemical and biological situations blocking the proper disinfection, cleaning, and decontamination of the well environment. While many of these test results are useful, analysis of the well water should also be directed more to the chemistry of mineral precipitation and corrosion and the biology and chemistry of the environmental bacteria capable of exerting influence on the well system. Well systems operate by the ability of the aquifer to flow freely toward a well structure. Pore spaces, those areas between the natural formation of sands and gravels, as well as fractures and joints in consolidated formations must remain open for the well to function properly. Bacterial growth and concurrent mineral accumulation within these areas effectively blocks water flow to the well.

Well blockage is usually a complex structure which in the situation of contaminant removal must be taken into consideration. The way the pathogen or chemical contaminant interacts with the structure is as important as the chemical reactions that take place or are used to remove the structure. Water analysis should be used to establish as closely as possible what is taking place within the well from both a chemical and biological position. These analytical results, together with observations of water quality, operational characteristics, and interpretation can be used to establish a profile of the well condition. Chemical parameters will indicate the type of minerals most likely present down hole. This could dictate some or all of the chemistry required for decontamination. Biological analysis will attempt to put a number on the bacteria present and will give some idea of the type and the volume of biofouling taking place. Bacterial identification can pinpoint specific areas of blockage or accumulation, the types of reactions that may or may not take place, and may indicate the method or best mechanical approach to use. When sampled properly, a water analysis can give a good indication of what condition the immediate aquifer is in and how flow through that area is affecting accumulation in or corrosion of the well. This observation would be helpful in assessing the potential interaction of contaminants in the well, and what chemistries, and in what sequence they would be most useful for contaminant removal.

Analysis Expressed as Calcium Carbonate

Certain parameters in water analysis are expressed as calcium carbonate or to be more specific as an equivalent quantity of calcium carbonate. The expression is usually used for hardness, its constituents calcium and magnesium, and the results of alkalinity titration, the phenolphthalein and the methyl-orange end points. The equivalent values are used to simplify comparison of analytical results as to possible precipitation of certain products. In the case of hardness and alkalinity we are looking for the potential reactions of calcium with the alkalinity ions (bicarbonates and carbonates).

For instance, if the actual ionic value of calcium were measured at 100 mg/l we would report it as 250 mg/l as calcium carbonate. The molecular weight of calcium is approximately 40 and the molecular weight of calcium carbonate is approximately 100. A comparison of the two shows a ratio of 1:2.5, hence the 250 mg/l.

Since values for "P" and "M" alkalinity and the hardness ions are those parameters that are reported in this manner, it allows us to assume theoretically but practical values for the bicarbonate HCO_3^- and the carbonate $CO_3^=$ ions. Having these values allows the chemist to determine the possibility of carbonate and non-carbonate precipitation in a water sample.

Recommended Chemical Tests

Table 8.1 is a list of primarily inorganic chemical parameters that are useful in establishing the mineral or chemical profile and in selecting the proper chemistry for well cleaning.

Table 8.1. Laboratory analysis for mineralization potential

Phenolphthalein Alkalinity*	Iron Ferrous (as Fe_{2+})
Total Alkalinity*	Iron Total (as Fe)
Hydroxide Alkalinity	Copper (as Cu)
Carbonate Alkalinity	Tannin/Lignin
Bicarbonate Alkalinity	Nitrate (Nitrogen)
pH Value	Sulfate (as SO_4)
Chlorides (as Cl)	Silica (as SiO_2)
Total Dissolved Solids	Manganese (as M_n)
Conductivity (micromhos)	Saturation Index
Total Hardness*	Chlorine (as Cl)
Calcium*	Potassium (as K)
Magnesium*	Sodium (as Na)
Phosphate (as PO_4)	ORP

*(as $CaCO_3$)

Alkalinity

Alkalinity measurements, together with the readings for hardness, calcium, magnesium, and iron, dictate the possibility for calcium and iron carbonate formation in the well. They are used for determining the Saturation Index described in Chapter 6. If the potential for carbonate precipitation is high, then additional acid will be required since calcium carbonate (calcite) will neutralize large quantities of mineral acid. Typical nomenclature used on analytical test results are "M" or "Total Alkalinity" or an older designation "Methyl Orange Alkalinity." The value given for any one of these designations is in mg/l as calcium carbonate, the value required for calculating the Saturation Index.

pH Value

The pH value will tell you whether a water is acid or alkaline. It will also give an idea of the corrosive condition of water (water below a pH of 7 almost always exhibit some degree of corrosive behavior). The pH is also used in the Saturation Index calculation and is an indicator of conditions effecting many reactions, both chemical and biological, within the well.

Chlorides

This analysis is often used to determine major changes in the water since the concentration of this ion doesn't usually change in a water supply. When there is mixing of waters often there is a change in the chloride level. It can also be used to pick up traces of other chemistry such as residual hydrochloric acid, sodium chloride (salt), etc.

Total Dissolved Solids (TDS) and/or Conductivity

The TDS reading, again a component of the Saturation Index, is also used to assess changes in aquifer water such as the dissolution of minerals which will increase the TDS of the well water. Changes in TDS usually verify recorded changes in specific ions such as the precipitation of calcium carbonate which would result in a reduction of the calcium in solution. Dissolution of iron pyrite in an aquifer would increase the concentration of iron and sulfide resulting in a corresponding increase in the total dissolved solids concentration.

Hardness

Hardness readings provide an idea of the overall potential for calcium carbonate, calcium sulfate, and magnesium hydroxide deposits. This is a reflection of hardness readings which are made up of the calcium, magnesium, and the iron concentrations. Values are given as mg/l calcium carbonate and are indicative of the interaction between these values and the total alkalinity, especially in the formation of the prior listed solids. If the total alkalinity reading exceeds the hardness level, then all the hardness is temporary and precipitation will be predominately calcium carbonate. If hardness exceeds total alkalinity then the difference is non-carbonate or permanent hardness. Non-carbonate hardness is usually representative of calcium sulfate or gypsum and you will most likely notice a substantial presence of sulfate in the analysis and might expect some gypsum deposits blocking the flow pathways.

Calcium

Calcium is part of the hardness reading, a component of the Saturation Index and a constituent of both calcium carbonate and calcium sulfate. Knowing the level will help predict the possible build up of either or both of these mineral deposits. In aquifers that have predominant dolomite formations the calcium concentration will equal the magnesium level or about 50% of the total hardness reading. In mixed or limestone aquifers the calcium concentration usually represents nearly two-thirds of the total hardness.

Magnesium

This metal is the other primary portion of the hardness reading. It usually does not form a carbonate mineral within the well unless dolomite is present in the surrounding aquifer. Magnesium will form a hydroxide, however, if certain conditions are met such as the dolomitic relationship of calcium to magnesium ratio of 1:1 and some hydroxide alkalinity present.

Sodium and Potassium

Both of these metals are usually constituents of very soluble minerals. The primary reason for their analysis is to balance the anionic and the cationic portion of the analysis as a method of verifying the analytical work. The sodium is also used to assess the presence of sodium chloride (salt) in the well water in conjunction with the chloride analysis.

Phosphate

Phosphates are often used in well work as cleaners and in well development for silts and sands removal. In addition there are some natural lithologies which contain deposits of phosphates. Analysis is used to assess the presence of phosphates and the potential of deposition in the gravel pack and other sections of the well. The efficiency of phosphate removal following use is also assessed as phosphates, which can be a stimulant for bacterial growth.

Iron as both Ferrous (Fe^2) and Ferric (Fe^3)

Iron can be both a natural constituent of the aquifer water and a by-product of corrosion. Since it participates in many oxidation and reduction reactions within the well, the types of deposits and their cause may be determined by an analysis. The presence of certain bacteria and other chemical parameters can also give indication of the type of iron deposits to expect.

Manganese

This metal, which behaves much like iron, can account for considerable deposits within a well environment. It can result in black deposits and a metallic taste to the water. Aquifer concentrations above 0.1 mg/l can result in significant accumulation resulting in well blockage if specific oxidizing bacteria or other oxidation activity is present.

Copper

This metal is determined both to balance the ion analysis and as an indicator of possible heavy or dense biofilm production. Copper can act as a primary linkage in biofilm structure and can increase its tenacity.

Nitrates, Nitrites, Sulfites

All of these are read to determine the mobility of ions in the aquifer and their potential involvement in deposit formation or bacterial growth. All members of this group may be considered both by-products of and an energy source for bacterial growth. In addition they may be contaminants from agricultural and industrial activity.

Tannins/Lignins

Analysis is made for these natural occurring products as an indication of surface water infiltration. Tannins/lignins are naturally leached products of leaves and other plant material. Unfortunately, certain aquifers contain very old deposits of these products so this possibility must be taken into consideration.

Sulfates

Gypsum (calcium sulfate) is one of the harder deposits to remove from a well environment and assessment of sulfate levels will give a good indication of its presence and the extent or its proportion to the rest of the deposit. Sulfates are also utilized by sulfate-reducing bacteria as an energy source. The presence of a high concentration or a change in the concentration in the aquifer water may indicate the presence of certain bacteria and the possible cause of corrosion, deposits, taste, or odor problems.

Silica

Silica, in rare circumstances, can deposit on screen and more often is seen as a portion of a deposit. The presence of silica in the blockage will require extra time and chemical consideration. Experience has shown that the silica content must be at or near 40 mg/l before significant silica deposits occur.

Saturation Index

The calculation of this index will help in determining the potential for calcium carbonate deposits and give good indication for gypsum formation. A negative reading may also indicate a corrosive condition within the well. The five parameters necessary for calculation of the index are: pH, temperature, total dissolved solids, alkalinity, and calcium concentration (see Tables 6.1 through 6.4).

Chlorine

This parameter is used to check on the validity of the sample, particularly if you are doing bacterial work. It also can be an indicator of insufficient flush out following disinfection.

Oxidation Reduction Potential (ORP)

The ORP reading is useful to determine what reactions will dominate or are possible in the well. Bacterial growth, particularly aerobic bacteria, appears to grow extensively forming dominant biofilms at the redox fringe of 50 to 150 mV (Cullimore, 1993). This level marks the transition area from the aerobic zone to the anaerobic level. A positive reading indicates an oxidative condition and a negative reading indicates a reductive chemical environment. Examples of a reducing environment is the formation of hydrogen sulfide gas or the precipitation of iron sulfides, the jet black deposits found deep in the well.

Recommended Microbiological Parameters

I have used the term parameters because the types of testing procedures used in microbiology investigation are not as well defined as in the chemistry laboratory. The results of two separate test procedures may be similar but not give comparative values. The values may be acceptable or usable as indicators of bacterial activity, but they must be understood and used with limitations. Considerations should include the test and the way the test is performed and recorded. Nevertheless, bacterial results are very useful in determining the condition of the well.

The first step in assessing the biological condition is to test for the overall bacterial count. This will place a load assessment on the well or serve as a yardstick for evaluating how much fouling is present. Table 8.2 shows the recommended bacterial testing for a profile of bacteriological activity in a well.

Table 8.2. Recommended microbiological profile for well water

Plate Count (colonies/ml)
Sulfate-Reducing Bacteria
Anaerobic Growth
ATP (cells per ml)
Bacterial Identification
Microscopic

Heterotrophic Plate Count (HPC)

Bacterial counts are not an easy determination if you take into consideration that anywhere from 90 to 99.9% are not culturable in the laboratory. The heterotrophic plate count is not much better as it only tracks aerobic heterotrophic organisms. It, however, is used extensively to track the bacterial activity in water systems, and well systems are no different. Therefore, its continued use is advisable because it allows comparison to much of the known and published data on well water analysis.

Have the analyst tell you what the range of their particular method is: 0 to 300, 0 to 3,000, or 0 to 5,000 cells per milliliter and use that method consistently. In this way you will come to know what a heavily fouled well versus a lightly or even a moderately fouled well will show on this test. You can also use the test, if used consistently and uniformly, over a period of time to assess whether a well is beginning to foul. This recommendation, I admit, is an oversimplification and many microbiologists will have objections and rightly so, but when used consistently with observation of general well conditions and water quality parameters, you will often pick up problems before severe well blockage or production loss occurs.

Adenosine Triphosphate (ATP) Analysis

Since a very limited number of bacteria are actually culturable, other methods have been used to assess the bacterial load on a system. A method used to count bacteria electronically results in a number which includes dead bacteria and particulate matter.

Another very useful method in assessing the bulk load of bacteria in a system is the cells per milliliter count determined by analyzing for the adenosine triphosphate levels in a given sample of water. ATP is released into a sample when the cell wall of the bacteria present has been destroyed or lysed during the testing procedure. Since the ATP is only present for about 15 seconds after a cell dies, only live bacteria are counted. Since all bacteria have ATP present all bacteria will be counted, both the aerobic and the anaerobic. (The Heterotrophic Plate counts only aerobic bacteria.)

Other advantages of this method are the relative short time to run the test, less than 15 minutes, and its excellent repeatability which together allow accurate growth and viability studies of the organisms. Since the ATP method essentially counts all the bacteria, whether culturable or not, be prepared for high counts. We have always converted the adenosine triphosphate value into the most probable cell count per milliliter based on the dominant organism size as viewed by light microscopy. This is done since most workers in the field are comfortable with the use of cell counts in assessing bacterial involvement having used the plate counts for many years. Other researchers, especially in Europe, often report the amount of adenosine triphosphate as μg/ml.

Sulfate-Reducing Bacteria (SRB)

Sulfate-reducing bacteria are relatively easy to evaluate and are an excellent indication of the severity of anaerobic growth in a well. SRBs are responsible for hydrogen sulfide production and the resulting taste and odor problems often seen in wells. As an anaerobic bacterium, SRBs exist in areas without oxygen usually found deep in a well or gravel pack, and/or under clay lenses, or any location protected from oxygen penetration. The hydrogen sulfide gas is forced upward producing a lowering of the pH or acid condition. This corrosive effect often causes severe damage to the well screen, the casing, and the drop piping. In addition to the damage or potential damage and water quality loss, detection of moderate to high levels will dictate a primary area that needs to be addressed during the cleaning.

Anaerobic Bacteria

Most laboratories are not set up for this procedure, but if the laboratory can give some assessment of total anaerobic count as compared to the total bacterial count some indications of the severity of biological fouling can be derived. Most clean well systems have less than 10% of the planktonic or free-swimming bacteria that are anaerobic. We have found in over fourteen years of testing that well samples with over 20% anaerobic growth have sufficient biological fouling to affect water quality and production. Anaerobic growth often indicates the severity of bacterial blockage with more mature or severe blockage showing increased levels of anaerobic activity.

Microscopic Examination

Centrifuging a portion of the well water sample and then observing the residue or the separated material under the oil immersion lense of a light microscope, hopefully equipped with phase contrast objectives, will give you considerable information concerning the down hole conditions. Many of the iron oxidizing bacteria will be recognizable, even some of the sulfate-reducing bacteria can be seen, albeit not too easily, especially to the untrained eye. In addition you will be able to assess the quantity in very broad terms of general bacteria present. Special bacteria such as branching type or heavy stalked bacteria can be observed and some comparative observation can be made as to whether they should be considered in assessing the biofouling in the well. Iron oxide (fluff) can be assessed and the presence of silica crystals in high numbers will often flag a severe silt and sand intrusion problem or one that is just starting. The presence of high concentrations of protozoans often signal surface water infiltration, especially if some chlorophyll containing ciliates are seen. Several times a high count of rotifers has confirmed a wastewater contamination.

Bacterial Identification

Many well professionals don't believe in spending time and money on identifying the bacteria that make up the larger populations present in the well water. This is unfortunate as periodically this information is very effective in explaining the primary cause of the well blockage and the resulting production loss. More importantly when the information is used to focus the cleaning procedures, time and effort can be very effectively directed at removing the blockage where a more random approach may have been less than successful. In recent years much information has been derived from research concerning the development of biofilms. It is now understood that the presence of different bacteria can produce tougher or denser biofilms, which chemistries can effect the biofilms, and where the different bacteria would be expected to produce the most dense formation or blockage. If a well cleaning or rehabilitation effort only cost ten thousand dollars the relatively minimal investment in bacterial identification could be considered cheap insurance.

It would be difficult to record all the instances in which bacterial identification can be of help in defining the blockage problem. Each well is unique and interpretation of the data should be left to the expert involved. Here are a few examples that may convince the unbelievers or at least provide for consideration of other possibilities.

Many of the iron oxidizing bacteria, the *Gallionella, Crenothrix, Leptothrix,* and the *Sphaerotilus* form distinct sheaths and deposit iron and manganese oxides in or on those sheaths. These aerobic bacteria almost always inhabit the screen areas or flow entrances to the well because of their need for a relatively continuous source of iron and the source of oxygen available as the water enters the well proper. The sheaths or stalks produced by the bacteria further complicate the area by bridging these openings. The iron oxyhydroxide deposits they form eventually dehydrate and the much harder or more dense ferric oxide is left. This deposit requires and responds well to brushing prior to chemical cleaning. If the well is an open borehole construction, the borehole wall will be the target of this matted architecture and some form of mechanical cleaning that will not harm the formation, such as jetting, often makes a difference in the success or failure of a chemical cleaning.

The slime formers are ubiquitous throughout the well but appear to produce the more stable biofilms in the gravel pack or the immediate formation outside the borehole wall. Knowing the relative strength and density of the blockage will dictate not only the chemistry but also how much effort should be put into the down hole cleaning.

Bacteria like *Pseudomonas aeruginosa* are well-known slime formers and are found in many wells. Since they are prolific growers, the bulk count does not give a good clue as to the severity of the blockage and we often see considerable variations in their presence over relatively short time intervals. Here, as with many similar bacterial biofilms, we rely on the presence of certain trigger organisms, which when present along with the *Pseudomonas* synergizes biofilm more than double the density and tenacity of biofilm produced by either organism alone. Work reported over the past few years indicate that the organism *Acinetobacter,* when present with *Pseudomonas aeruginosa,* produces just such a tough biofilm

(Molin, 2000). Cleaning, rehabilitation, or even decontamination efforts must be tailored to account for this type of growth and both mechanical and chemical efforts should be adjusted.

Research done at Umea University in Sweden and other laboratories in England and Germany and used in our own laboratory over the last five years, indicates that certain morphology exhibited during colony formation, known as the rugose characteristic, is indicative of the capability of more severe biofilm formation or production. In addition, in a more recent publication researchers have attributed this same expressed characteristic as indicative of increased resistance to chlorine treatment (Jass and Wai, 2000). Isn't this type of information important when designing or specifying the cleaning effort?

A considerable amount of the biofilm research conducted in the food processing and medical field has application in the well environment. These findings have been reconstructed and observed in our own laboratory. Not only are the phenomena related here seen in mock well cultures in the lab, but they have been observed in the field by correlating laboratory analysis with the observations and readings taken of specific wells.

Field Testing of Deposits

While analysis of the well water is far more important in determining the type of blockage that can be present in a plugged well, occasionally samples of deposits are obtained. Without a complete field laboratory the analysis is usually restricted to the results of acid dissolution of the incrustation.

Observation should first be made as to color and density of the sample.

- ***Black deposit*:** may indicate an iron sulfide or a manganese deposit.
- ***Dark to reddish brown*:** usually indicates a ferric iron oxide deposit.

- ***Bright yellow***: most probably sulfur and usually seen high up on the drop piping and casing, often above the water level.
- ***Light tan deposit***: can be dolomite, a mixture of calcium and magnesium carbonate.
- ***Very light color to white***: calcium carbonate, usually seen with other minerals providing additional colors.
- ***Very heavy or dense deposit***: the deposit is predominately mineral.
- ***Very light or low density deposit***: Considerable biological or organic material present.

Placement of a few drops of hydrochloric acid or muriatic on the incrustation may elicit some additional information.

1) Considerable foaming or frothing will indicate a carbonate (calcium, magnesium, or iron).

2) Hydrogen sulfide gas or rotten egg odor indicates iron sulfide present.

3) A strong chlorine odor will indicate the possible presence of manganese dioxide.

4) No effervescent or frothing or odor usually indicates the presence of iron oxide, and/or calcium sulfate and/or silica.

There are a number of simple field tests which can check for iron, sulfate, calcium, and phosphates. If these are available dissolve a small amount of the deposit in hydrochloric acid using as little acid as possible. Dilute the dissolved material with deionized water to reduce the acid strength and to give sufficient volume for testing. Positive tests for any of the parameters listed above should confirm some of the observations made earlier.

CHAPTER 9 Selecting the Chemistry and Its Application

As with any cleaning operation proper selection of chemistry and its application is of paramount importance. Careful consideration must be given to selection and use of chemistry as well as its application because of the possible uniqueness of the contaminant and certainly the area of operation, the well. Contaminants will require dissolution, suspension, and neutralization and each of these reactions will require either acid or alkaline pH changes along with either oxidative or reductive conditions. Since the complexity of the well environment will incorporate chemical reactions both of and against the microstructure (which includes the normal mineral and biological structure of wells), correct scheduling of the programmed cleaning is a necessity. Caustic, used for destruction of a toxic chemical or clean up of hydrocarbon contamination, could result in excessive precipitation which might block further recovery of the contaminant. Chlorine or oxidizing agents may react with heavy biofilm deposits that would block the removal of pathogenic bacteria or chemical contaminants. Hence, it is important to understand the complexity of the well system and to account for it when selecting the proper chemistry.

The Well Environment and Volume Calculations

During the cleaning or decontamination of any piece of equipment or area the parameters of that area must be calculated. Surfaces to be cleaned and volumes of cleaner required must have a uniform method of calculation. Therefore, in order to develop a good procedure for decontamination, disinfection, or chemical rehabilitation the parameters must be established. Since the volumetric consideration should include all the surfaces within the well that need to be cleaned, the area which includes the interior volume of the well represented by the standing well volume (SWV) and the area which includes the gravel pack (less the space taken up by the pack material) should be calculated. In actual practice, factors of 1.5 times the SWV for wells with casing diameters greater than six inches and two times the SWV for wells with casing diameters six inches or less are used to determine the volume of cleaning chemistry required. This volume should reach all areas of the well system that are in need of cleaning including structural surfaces and the microstructures or flow pathways of the well. Most of the time the calculated volume provides for some penetration of the immediate formation which is an excellent practice. Figure 9.1 shows in visual form parameters which should be used on a typical gravel packed well.

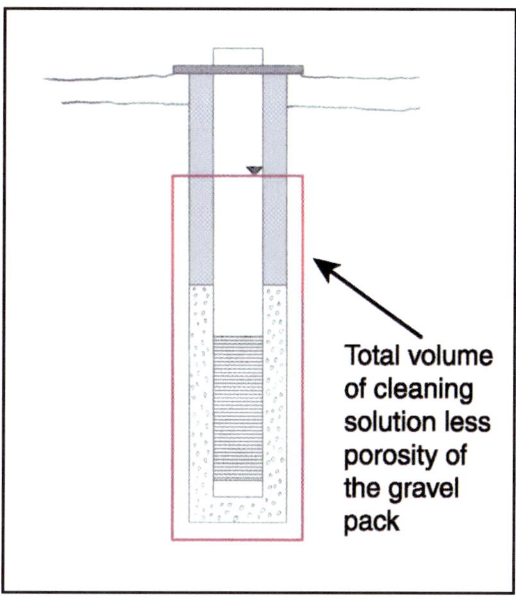

Figure 9.1. Target cleaning area of typical gravel packed well

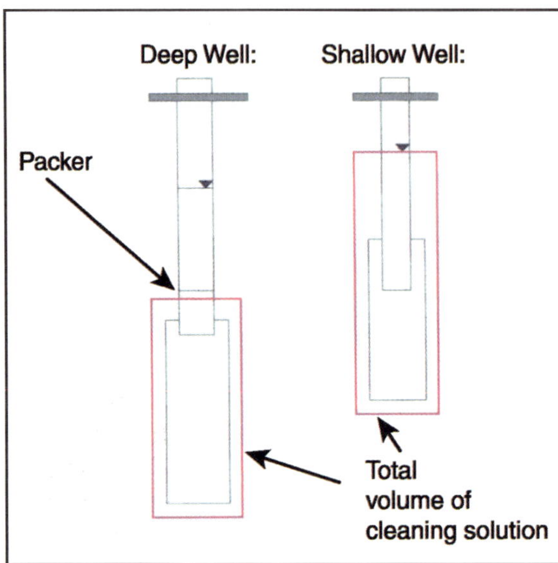

Figure 9.2. Target cleaning area of typical open borehole wells

I recommend a similar practice for open borehole well systems (see Figure 9.2) as these wells are in need of a chemical volume sufficient to penetrate the formation more extensively. This will insure absorption of the greater concentration of dissolved and suspended solids often released during cleaning of this type of well construction. An additional consideration necessary when cleaning an open borehole well is the potential for lost solution in the many fractures which open into the well proper of consolidated formations. Often these fissures lead from the open borehole to other openings in the sandstone or carbonate formations (see Figure 9.3).

Of course there are additional constraints that should be taken into consideration with other types of well design and in view of the cleaning method or procedures utilized. In some areas of the country well design often includes a substantial underreaming. This results in an extensive gravel packed area in the bottom of the well. Since this area is usually heavily impacted with biological and/or mineral blockage, an additional volume of cleaning chemistry will be required.

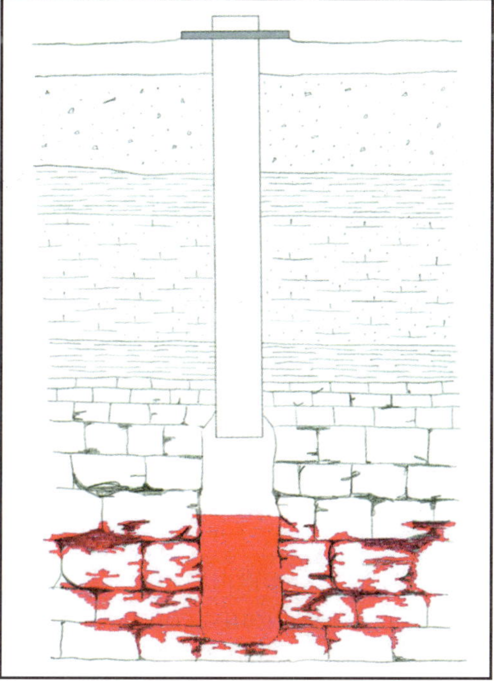

Figure 9.3. Open borehole well showing chemical loss

Additionally, wells that are very deep with screen placement in the lower areas of the well are often cleaned by isolation of those areas with packers (see Figure 9.4). The cleaning chemistry calculated for that area is then injected through the packer and isolated at least within the casing and screen. Surging or other mechanical activity will also have to be applied through the packer to maintain the isolation of the chemistry within that area. Unfortunately this practice is not always that complete. Often the chemistry is tremied into a zone and the mechanical application carried out oblivious to the rapid dilution of the cleaner. Some degree of cleaning can be achieved by calculating sufficient chemistry to prevent unwarranted dilution. This is achieved by calculating the volume contained in the section of screen to be cleaned. Multiply this figure by three. This will provide sufficient chemistry to clean the gravel pack and immediate formation in the area required by supplying equal volumes of chemistry above and some below the area to limit dissolution. Injection should begin above the area to be cleaned at a distance equal to the section to be cleaned and released evenly throughout that section and the entire section of screen to be cleaned. Surging during the procedure will draw on the chemistry above the screen and lifting of the surge block will draw on chemistry which has gravitated to the lower levels.

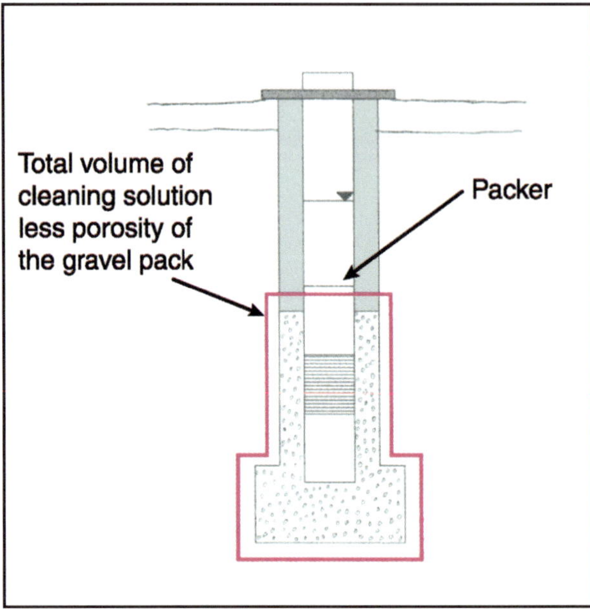

Figure 9.4. Large underreamed well using packer placement

Another method of isolated cleaning in deep wells involves the use of a double disk surge block through which chemistry is tremied. This procedure often has some failure that can be attributed to insufficient cleaning chemistry. Again it involves determining approximately just how much dilution will take place and accounting for that dilution by the addition of more chemistry. If the surge block is ten feet long and the area to be cleaned is 100 feet, sufficient chemistry to clean the 100 feet of both screen and gravel pack (1.5 to 2 x the area volume) must be used. The chemistry must be systematically added throughout the surging process. In this way some level of chemical savings is achieved and a strong cleaning chemistry is equally applied. If, however, the cleaner is added in the first pass of the surging, then sufficient dilution usually takes place to limit all but carbonate dissolution.

Selecting Proper Chemistry

Laboratory tests and possibly use of a video camera in the well should be utilized to identify the biological, chemical, and structural conditions in the well. Most likely the contaminant has already been identified; however, a complete biological and chemical analysis should be run on the well water in order to determine the degree that biology and mineral buildup will influence the cleaning procedure. Mineral deposits could require a considerable amount of acid and heavy biological involvement will require special consideration. Videoing will show the condition of the interior of the casing and screen. The physical condition, as well as the specific contaminant, must be taken into consideration in selecting the chemistry, the sequence, and method of application. Site conditions may also have some influence on the handling, storage, and use of certain chemicals. Sulfamic acid based products such as NuWell-100 and NuWell-110, both of Johnson Screens, Inc., are often the chemistry of choice in small well applications and in difficult terrain due to the ease of transportation and storage. Another product with a different acid base useful because of its powder formulation is UNICID Granular of Design Water Technologies.

Acids are used primarily for mineral removal during cleaning procedures and are a major and necessary constituent of the cleaning chemistry. While many wells do not accumulate hard water precipitates such as the carbonates or gypsum from non-carbonate hardness, they usually suffer from iron oxide accumulation from either corrosion of the well structure or oxidation of aquifer iron. The laboratory analysis will give a good indication if either or both situations exist. Carbonates, usually calcium, and sometimes iron and magnesium will require more acid as all carbonate based mineral formation will neutralize considerable amounts of acid. Iron oxide dissolution will perform much better with some type of chelating or sequestering agent available once the acid has dissolved the deposit. Manganese formation also requires a chelating or sequestering chemistry to improve removal. While chelating or sequestering acids like citric or hydroxyacetic have been used in the past, the newer polymers or biodispersants are many times more effective.

Acids will promote dissolution of most minerals when the pH drops below 7.0; however, dissolution moves much quicker at a pH of 1.0 to 2.0 especially for the more dense gypsum and iron oxides. The pH should be maintained at 3.0 or below during the entire time of the acid cleaning. This is important since solids will fall out of solution (precipitate) very quickly above a pH of 4.0 and thus considerable material will be left in the well. Numerous laboratory tests have shown that more than 50% of the solids fall out of solution or begin to precipitate as the pH rises between 4.5 and 5.0.

In well systems that do not show a potential for carbonate, sulfate or oxide deposits, acid is still a good choice as the primary cleaner although the concentration should be limited. Once a pH of 1.0 to 2.0 is achieved the well should be worked for a short period of time and then monitored again for pH. If there hasn't been a significant rise in pH then usually no additional acid is required. Frequently in this situation the initial acid charge is approximately 3% by weight of the cleaning solution volume.

A secondary consideration would be the condition of the solution. Heavily loaded cleaning solutions (high dissolved and suspended solids content) could greatly reduce cleaning even though pH remains low, and in addition may result in movement of solids further out into the formation. Solutions that have a conductance greater than 40,000 μS/cm and a turbidity >300 NTU fall into this classification. In these conditions, as with any cleaning, due consideration should be given to evacuating the cleaning chemistry and starting over. Very seldom will the addition of more acid and dispersants or other additives help the cleaning procedure.

Hydrochloric Acid

Table 9.1 is a chart of various acids often used in well cleaning. Hydrochloric is by far the most used cleaner and generally provides quicker dissolution of many of the mineral deposits found in wells. This acid is also a more universal solvent as it tends to be a more simple acid and does not provide conflicting "common" ions. An example (Equation 9.1) is the use of hydrochloric for the dissolution of gypsum or calcium sulfate.

Equation 9.1.

$$1)\ CaSO_4 + 2HCl \rightarrow 2H^{++} + SO^{=}_4 + Ca^{++} + 2Cl^{-}$$

Equation 9.2.

$$2)\ CaSO_4 + H_2SO4 \rightarrow 2H^{+} + 2SO^{=}_4 + Ca^{++}$$

Since neither calcium nor sulfate are constituents of hydrochloric acid, there is no net increase in any of the ions contained in the reaction. This not only promotes more dissolution but limits reprecipitation during the cleaning procedure. Equation 9.2 shows the use of sulfuric acid indicating a doubling of the sulfate ion which encourages reprecipitation of sulfates and/or limits dissolution of the gypsum. Sulfuric acid was used as the example since sulfamic acid is probably the second most used acid for well cleaning. Initially the sulfamate ion formed is very soluble and no reprecipitation of calcium sulfate takes place, however after only two to six hours of use the sulfamic acid is hydrolyzed to form sulfuric acid in the solution. This significant increase in the sulfate ion from either the sulfuric acid or the hydrolysis of sulfamic acid would limit further cleaning and could result in the reprecipitation of calcium sulfate further out into the gravel pack or formation surrounding the well.

Table 9.1. Characteristics of common well cleaning acids

Characteristic	Sulfamic	Hydrochloric	Phosphoric	Hydroxyacetic	Citric
Appearance	White crystal	Slight yellow liquid	Clear liquid	Clear liquid	White crystal
Formula	HSO_3NH_2	HCl	H_3PO_4	$CH_2OHCOOH$	$C_6H_8O_7$
Molecular Weight	97.1	36.47	98.0	76.05	192.12
Type	Mineral	Mineral	Mineral	Organic	Organic
Hazardous Fumes	None	High	None	Some	None
Relative Strength	Strong	Strong	Strong	Weak	Weak
pH @1%	1.2	0.6	1.5	2.33	2.6
Relative Reaction Time 1 = Fast 10 = Slow	<2	1	4 – 5	4 – 5	4 – 5
Corrosiveness to:					
• Metals	Moderate	Very high	Slight	Slight	Slight
• Skin	Moderate	Severe	Moderate	Slight	Slight
Reactivity vs:					
• Carbonate Scale	Very good	Very good	Very good	Poor-fair	Poor
• Sulfate Scale	Good	Good-poor	Good-poor	Very poor	Very poor
• Fe/Mn Oxides	Fair	Very good	Good	Good	Chelates
• Biofilm	Poor	Poor	Poor	Moderately Good	Poor
Pounds of Acid (100%) Required to Dissolve 1 lb. of Calcium Carbonate	2.0	.73	.65	4.5	4.0

Some of the major problems in the use of hydrochloric is its high corrosion rates against stainless and low carbon steels and the limited access to food grade or NSF certified products.

Sulfamic Acid

While use of sulfamic acid is limited against gypsum or sulfate containing blockage, it is a useful acid for smaller wells and on any other types of deposit. Sulfamic acid is also more conveniently transported due to its crystal and/or powder form. Due to the convenience, this acid is often used where iron oxides are a problem; however, the acid is not particularly effective against iron deposits. A chelating or dispersing agent should be used to increase solubility of the iron. An old trick often used is to add salt, sodium chloride, to the cleaning solution. This will convert to form some hydrochloric acid which is much more effective against oxides of iron and manganese. This, however, greatly increases the corrosion effect against stainless steel since hydrochloric acid or the chloride ion is considered very corrosive toward chrome alloys of iron. Usually when hydrochloric acid is used an inhibited product is purchased or a separate inhibitor is added. The limited life of the inhibitor during acid cleaning (four to eight hours) is usually overlooked and corrosion of well structures can be severe at times.

Phosphoric Acid

Phosphoric acid is an excellent replacement for hydrochloric in well cleaning applications. It can be easily purchased in a food grade quality acid. Phosphoric also has very limited corrosive activity and actually provides some passivation of the well structure offering some future corrosion protection. The acid is tribasic, meaning that all of the hydrogen ions are not available immediately, making it a much slower reactive acid than hydrochloric. This can be used to advantage by producing a much slower release of carbon dioxide from carbonate dissolution improving the safety of the operation considerably. The acid is also much safer to handle since it lacks the dangerous hydrochloric acid fumes and is much less corrosive to skin, eyes, etc. Phosphoric has had limited use because of the possible production of phosphates in the formation and the resulting stimulus for bacterial growth. Studies using the Johnson Screens NuWell-310 product, a polymeric biodispersant, have shown that as long as a strong dispersant is used during the cleaning procedure, little or no phosphate formation occurs and those phosphates formed are maintained in solution by the dispersant and easily removed during pump out. The chelating or sequestering effect of the phosphoric also improves clean up of metal surfaces down hole increasing iron dissolution and removal. In addition to laboratory research hundreds of water samples have been analyzed over the past ten years from field applications where dispersants were used, no significant phosphate residue was found. The dispersants used successfully were Johnson Screens' NuWell-310, Layne Christensen's QC-21, and Design Water Technologies' UNICID Catalyst and E.P.A.C.

Organic Acid

Of the organic acids, hydroxyacetic (glycolic) and citric are the most often used. Hydroxyacetic does provide some activity against biofilm removal, however the solubility of some of the by-products formed with calcium is limited and often results in some plugging of the formation. In addition, this acid can be a carbon source for bacterial growth. Care must be taken to assure complete removal from the gravel pack and formation area. To an even greater extent this is true of citric acid which has an abundance of carbon molecules in its makeup which can result in a rich energy source for bacterial regrowth.

Acetic acid is another organic acid which is used occasionally for well cleaning as well as pH control during chlorination. Most of it used for the latter is in the form of vinegar which is a 5% solution. In this form it is relatively benign and the very low concentration used has limited effect on encouraging bacterial growth. Glacial acetic, the highly concentrated form, is used more as an acid cleaner. Glacial acetic is a difficult acid to use due to its pungent vapors which can be dangerous to both the lungs and skin producing severe tissue destruction.

While use of the acid has certain safety drawbacks the acid effect on calcite and gypsum formations can result in the production of by-products with limited solubility particularly in the fringe areas of the cleaning zone. Acetic, like hydroxyacetic or glycolic acid, also provides an excellent carbon source for bacteria.

Caustics

Caustic or alkaline chemicals are not often used in well cleaning but can be used and indeed are the chemical of choice in some unique situations. Strong bases such as potassium and sodium hydroxide are usually the products of choice, however alkaline phosphates and soda ash have also been used. The situations for which they have been employed most often are oil contamination of the well environment and excessive biofouling. Caustics saponify oils and as such make them water soluble or at least miscible in water allowing the oil to be washed from the area of contamination. Caustic tends to break down the exopolymeric material and, as in the case of the oils, solubilize it making it more easily removed from the formation, gravel packs, and well structure.

Alkaline chemistry is carried out at pH values above 7.0 and more often above a pH of 11. This is difficult chemistry in a mineral environment since many mineral compounds readily precipitate at pHs above 8.0 and some begin precipitation above a pH of 5. With the development of the strong biodispersants, such as NuWell-310 by Johnson Screens, very little caustic cleaning of any water system is performed for the purpose of biofouling removal since the dispersants are so effective and have none of the induced precipitation problems. Oil contamination is still more effectively treated with an alkaline cleaning solution utilizing one of the strong bases and a surfactant to maintain control of the fringe areas of the cleaning and to improve penetration of the gravel pack and formation. Alkaline cleaning may also be helpful in decontamination procedures where an alkaline reaction is required to neutralize or destroy a pesticide or toxic chemical. If this is required, some of the same techniques that are used to alleviate the precipitation problem in oil removal could be employed for a decontamination project.

As reviewed in Chapter 6 the precipitation of calcium carbonate and calcium sulfate increase as the pH and alkalinity rises. Iron and manganese will also oxidize more readily with some increase in pH. In order to control the possibility of precipitation during alkaline cleaning, some provision for ion control must be made to prevent the formation of the precipitate. Strong dispersants have been shown capable of controlling precipitation if provisions are made to control crystal formation when dispersant control breaks down or its capacity is exceeded. Chemistry to prevent crystal formation is a secondary control feature. Due to the high probability of precipitation and the ease that dispersant capabilities are exceeded under high alkaline conditions, it is a necessary protocol. If an acid cleaning is used to pickup any precipitates remaining after flush out of the alkaline cleaner, use of a dispersant at this time will also improve removal.

When caustics are used in a mineral environment greater than 90% of all the solids are lost from the solution as the pH is pushed past the 10.0 level and a pH above 12.0 is required for most cleaning projects. If dispersants are used and properly designed such as the Johnson Screens NuWell-320 which incorporates not only dispersency but also crystal modification and suspension polymers, then loss of precipitates can be held to a 10 to 15% level. A follow-up acid cleaning will easily remove these deposits satisfactorily and prevent plugging especially of the formation. Acid cleaning will not supply the solubilizing effect against oils necessary to clean the aquifer and should not be used without first removing an oil based contaminant.

As mentioned above both sodium and potassium hydroxide are used for caustic cleaning, however potassium is the chemical of choice. The potassium salt is more soluble and will create less opportunity to induce precipitation of sodium salts of which there are more in most geological formations. The product is cleaner and less corrosive to skin so there are some safety advantages. The two cleaning solutions most often used are 3 and 5%, with 5% the concentration of choice. This is especially true in areas where the oil contaminant has been in the well for an extended period and partial biodegradation has taken place and considerable biomass has developed along with the oil contaminant. The dispersant is usually used at a 3% concentration. Alkaline phosphates are not used for oil decontamination as much anymore since the role of phosphate stimulation of biogrowth has been observed and the chemistry of phosphate precipitation with tricalcium phosphate formation can be anticipated. Soda ash was never used extensively since the addition of carbonates into a well environment containing calcium easily precipitated calcium carbonate and further complicates the cleaning process. *All suggested use levels given are percent active product and should be calculated based on the total volume (weight of) of the cleaning solution (see Chapter 10).*

Alkaline Products Used to Neutralize Acid Discharge

Soda ash is used extensively along with lime (calcium oxide) and magnesium hydroxide as a neutralizing agent to prepare the acid pump out for discharge to a sewer system, etc. Sodium carbonate (soda ash) is a carbonate containing product and will release carbon dioxide upon neutralization. This release could cause splashing which might become a safety problem. Likewise the off-gassing of the carbon dioxide could further compromise the safety program if

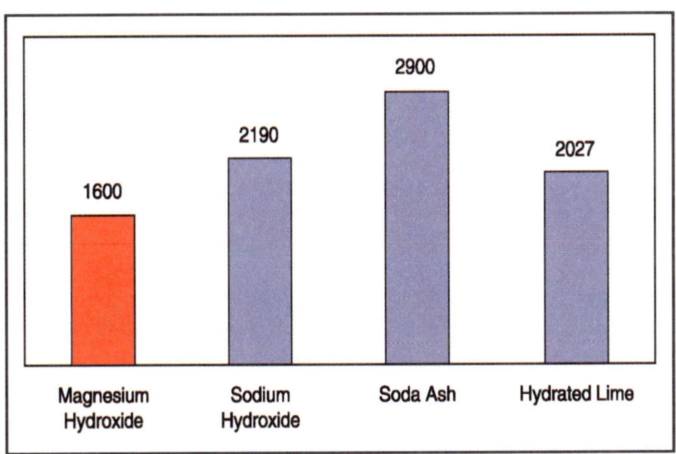

Figure 9.5. Pounds required to neutralize 1 ton of hydrochloric acid

neutralization took place in a confined area such as a vault or pit. Magnesium hydroxide has become a product of choice in recent years because of its ability to limit or reduce total dissolved solids in the discharge (see Figures 9.5 and 9.6). Furthermore, its liquid

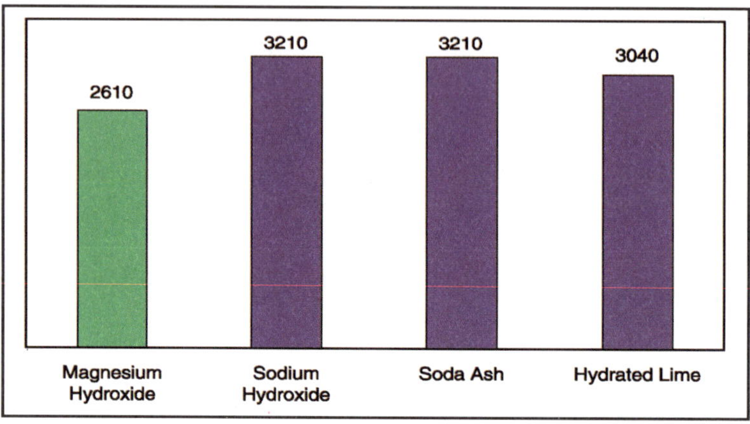

Figure 9.6. TDS in effluent per ton of hydrochloric acid neutralized

state (it is sold as a 50 – 60% slurry) as well as its buffering near a pH of 9.0 makes it an easy and safe product to use.

The buffering prevents discharge of excessively alkaline or high pH water. A unique product on the market called pH Neutralize manufactured by Design Water Technologies also prevents discharge of excessive solids and high pH water. This product has excellent dispersion activity allowing more precise neutralization and is often used on hazardous waste sites and other sensitive environments.

The liquid caustic, sodium hydroxide, is also used to neutralize acid discharge. The product has some good cost advantages as well as handling considerations. As a 50% active liquid it can be supplied in 55-gallon drums and easily pumped into a discharge tank for neutralization. The drawback, however, is that it is very easy to overextend the neutralization and produce a high pH. To prevent this most contractors using liquid caustic use the product to partially neutralize the low pH water and complete the process with soda ash. This will limit the pH to a maximum of 9.5.

Dispersant Polymers

For over sixty years well cleaning has revolved around the use of acid and chlorine to remove the mineral and biological masses that affected the waterflow. Mineral acids, as noted earlier, are used to

Figure 9.7. The congested well

dissolve the mineral blockage while chlorine has been used to control the biofouling. Modern chemistry uses polymeric dispersants to control solids during chemical cleaning and to provide dissolution of the biological portion of the blockage. This control not only increases the activity of the acid cleaner against the mineral/biological matrix but greatly improves pump out of the dissolved and suspended products.

Concentrated Cleaning Environment

In the highly concentrated well cleaning environment, mineral deposits are readily dissolved with the use of acids, particularly those of the carbonate salts which release the metal ions of calcium, magnesium, and iron into the solution as well as the carbonate and bicarbonate anions. While carbon dioxide is released from the solution as a gas, the concentration of metal ions remains. In addition to dissolved forms, partially solubilized crystalline structures of sulfates, oxides, and silicates also impact or crowd the solution. Also present are the ions of the acid itself and the suspended insoluble matter consisting of clay particles and bacterial aggregates comprised of the bacteria and the associated bacterial exopolymers (see Figure 9.7). It is no wonder that the acids were only partially successful as well cleaners. In the congested environment minerals precipitate almost as quickly as they are

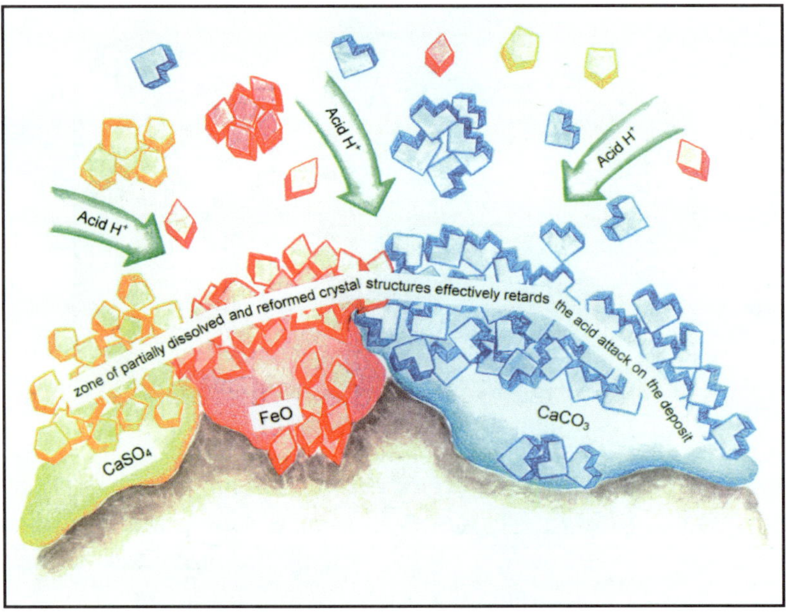

Figure 9.8. The slowing or blocking of acid cleaning

dissolved and this precipitation often takes place in the formation areas further out than the original blockage.

As the acid attack takes place on the surface of mineral deposits, the products of partial dissolution and reformation begin to accumulate on the surface blocking the acid movement toward the mineral deposit (see Figure 9.8). This is the major reason why agitation is required to help remove partially dissolved structures so that the acid can continue to dissolve the more readily soluble mineral compounds. The well environment however is isolated and it is difficult to circulate cleaning chemistry through or over all the surfaces, and much of the cleaning activity must take place away from the surging action. Therefore some chemical means of controlling the reprecipitating of solids is necessary. Dispersant chemistry offers this control.

Chemical Dispersion

The use of dispersant chemistry to enhance the reaction can greatly improve the efficiency of acid cleaning. "Dispersant chemistry is the use of specific polymers to block the attraction of positive and negative forces which account for the formation of

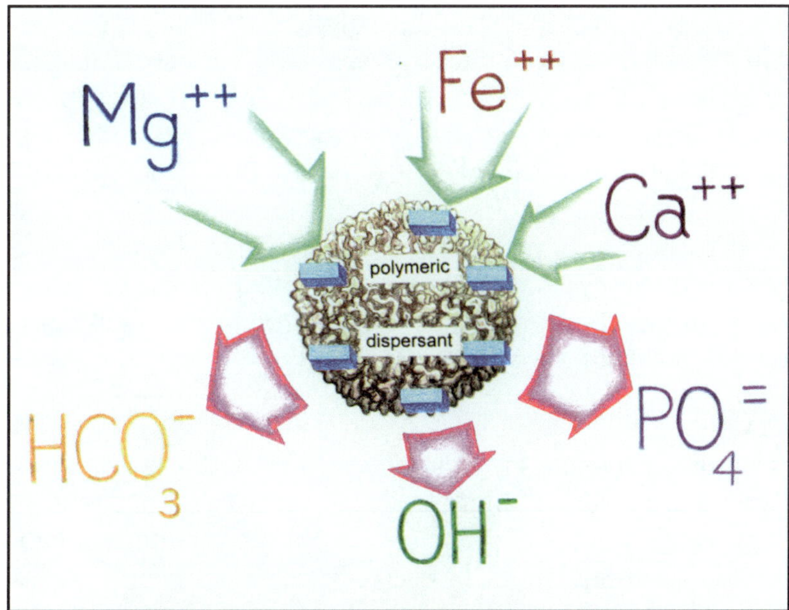

Figure 9.9. Effects of dispersant polymers on mineral ions

salts (compounds), crystals, granules (macrocrystals), and colloidal masses" (Schnieders J., 1996). All of these entities form and reform during the acid cleaning reaction and choke or slow down the chemical activity. Figure 9.9 is an example of a specific polymer used to disperse the charged dissolved and suspended products of the acid cleaning reaction. The polymer has a very high density negative charge and will repel negative charged particles or ions and draw to it ions or particles having a positive charge.

Salts (Compounds)

Complete dissolution of the crystal deposit releases ions into solution. (One of the simplest acid mineral reactions is that of acid against calcium carbonate which readily dissolves freeing calcium and the carbonate ions into the solution.) The calcium ion is positive while the carbonate ion is negative, and as these entities increase in concentration in the dissolution process, the random reformation of calcium carbonate is inevitable. While some of the carbonate ion is lost with the evolution of carbon dioxide, the calcium ion remains and other minerals such as gypsum contribute both calcium and sulfate to the solution. With specific anionic (negatively charged) polymers present, the cationic ion (positive charged) can be neutralized. In reality it is more of a push and shove situation since the cation is drawn close while the anion is repelled. That is why the term dispersant is so appropriate. This prevents the reattachment necessary for the reformation cycle and prevents formation of precipitates. The negative polymer control of the reformation cycle would apply to most metal salts such as sulfates, oxides, and more complex structures which normally makeup deposits in the well environment.

Crystals

If calcium carbonate growth continues it becomes a crystal large enough to drop out of solution. The crystal is a geometric placement of the molecules of the salt in such a way that the formation is repetitive resulting in a latticework type structure. A highly negative-charged polymer will wrap itself around a positive charged position on a crystal preventing its further growth. The negative side of the salt's molecule must attach itself to the positive position on the crystal for crystal growth to proceed. The blocking effect of the polymer is the crystal modification or growth control that gives the better "enhancer polymers" or "biodispersants," such as Johnson Screens' product NuWell-310 and NuWell-320, Layne's QC-21, or Catalyst by Design Water Technologies, a secondary control feature during the cleaning procedure. This reduction in the concentration of crystalline solids greatly increases the ability of the acid to solubilize the lesser soluble mineral compounds (i.e., sulfates, oxides, and phosphates).

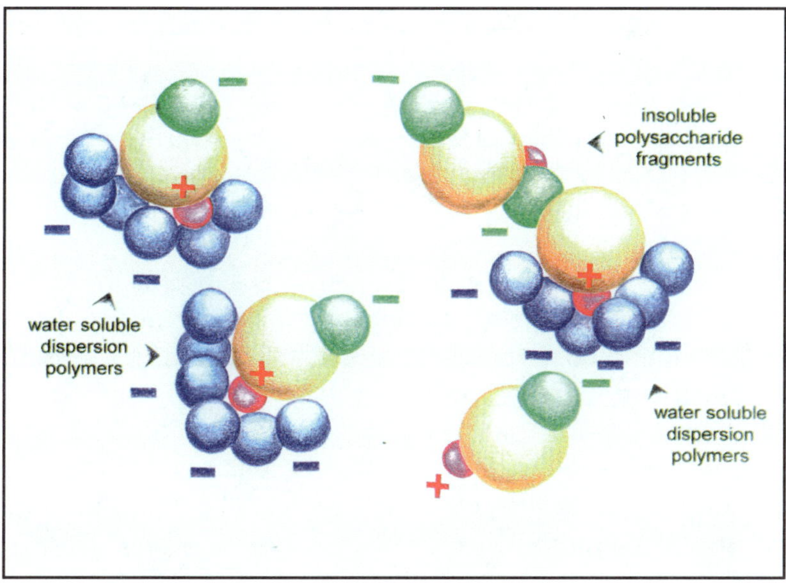

Figure 9.10. Effects of dispersant polymers on exopolymers
Dispersion polymer blocks reforming of the long chain polysaccharides and creates soluble colloids which are easily washed from the system.

Colloidal Masses

If chlorine or other strong oxidizing agents are not incorporated into the cleaning activity, the mechanical agitation of an acid solution alone will begin to partially break apart the biomass. Since 75 to 80% of the biomass is an insoluble polysaccharide polymer, this breakup results in the movement of fragments of the polysaccharide material into the free flowing water. Without dispersion chemistry this polymer material merely reattaches and reforms the saccharide polymer again on other surfaces or mends its own biofilm formation. The uniqueness of dispersion chemistry is such that the exposed areas of the fragmented polysaccharide polymer offer excellent attachment (positive sites) for the negative-charged dispersion polymers. This attachment results in control of the polysaccharide material by producing a molecule with a soluble side (anionic polymer) and insoluble polysaccharide portion assuring it of being maintained in solution or suspension (see Figure 9.10). In addition, the presence of the negative charged dispersion polymer prevents the reattachment of the saccharide monomer so that now the polymer is much shorter and less adhesive or attachable to the remaining polysaccharide film. This allows systematic removal of the biofilm from the system during flush out and is the reason the particular polymer is referred to as a biodispersant.

Dispersant Use

Solubility of both minerals and organics is substantially increased through the use of dispersion chemistry primarily due to its blocking effect on the force of attraction (Fowkes, 1965). As the surfaces of chemical entities (ions, molecules, crystals, polymers, or aggregates) become coated with a negative charge, the repulsion of similar charges results in maintaining these products in solution or suspending them in the cleaning liquid. The reaction prevents reprecipitation and agglomeration of particles toward larger sizes which otherwise would eventually settle and be difficult to remove from the well system. The use of the right dispersants will therefore increase the effectiveness of the pump out and result in three to four times more solids removal from the well (Research Investigation NuWell Products 310 and 320). Dispersant chemistry has been used in industrial applications for many years as polyelectrolyte control polymers in boilers and cooling systems for scale or deposit control. The use of polymers in severe acid conditions does require special configurations so that they are active at the low pH levels necessary for well cleaning.

Polyphosphates

Phosphates are used primarily as silt and clay dispersants during well development. In recent years the practice has become somewhat controversial as phosphates stimulate bacterial growth. Originally it was thought that the phosphate was removed during the pump out, however the interaction of phosphates with calcium usually produces insoluble salts which remain in the well often resulting in bacterial problems over extended periods. Phosphate chemicals are available in a wide range of products which are classified as an acid or alkaline product. Just which product is most effective in a particular area may be difficult to determine and has often been decided by trial and error over time. In areas of considerable calcium carbonate acidic phosphate products can cause dissolution of the formation with subsequent production of by-products of limited solubility resulting in blockage.

In recent years very active polymer solutions have been developed which have replaced phosphates especially in large well development. Because of their high activity their use is quite economical. They are usually referred to as mud busters or clay dispersants. While phosphate products are used at between 0.5 and 2.0% by weight of well volume, the polymer products, which are liquids, are used at 0.2 to 0.4% of well volume.

The active polymer products should not be confused with the products based on acid salts. These products are mostly granular products and depend on acidifying the solution to loosen or break down the mud. They are useful in breaking the mud formation but have very limited effect on removal. Commercially available products based on clay or mud active polymers are Design Water Technologies' Mud Buster or Johnson Screens' NuWell-220.

Surfactants

Surfactants or surface-active agents represent a wide range of chemistries which are designed to modify the surface tension of liquids against solids. They are used specifically in well cleaning to facilitate penetration of the gravel pack and formation as well as any deposits in the system. While there are many types of surfactants, those that are active over a wide pH range (amphoteric) are most useful as they can be used in both acid cleaning and cleaning in the alkaline pH range with caustics. Certainly low foaming would be an advantage due to the high probability of mechanical activity. A resistance to chlorine or oxidative chemistry would allow their use during chlorination. A primarily hydrophilic surfactant would provide better blending and prevent loss of the product in the aquifer through separation. Surfactants which are reasonably concentrated (at least 75% active) are usually used at between 0.2 and 0.5% concentration in the cleaning solution.

Chlorine Application

Chlorine compounds are used in well cleaning primarily as disinfectants and as biofilm removal products. As a biofilm removal agent they are only partially effective; however as a disinfectant, if used properly, they are very good. Chlorine is available as chlorine gas, as a liquid (sodium hypochlorite), as a solid (calcium hypochlorite), and as a gas (chlorine dioxide). Chlorine as a gas is used by a few contractors but most find it too dangerous and difficult to handle. Those using it have found it very effective as the gas imparts a pH adjustment to the water and as seen later the hypochlorous acid (which dominates at a slightly acid pH) is far more effective as a biocide than the hypochlorite ion. Chlorine dioxide, also a gas, must be generated on-site; and while it will penetrate the biofilm better than the hypochlorite, it does not maintain a residual thereby reducing its effectiveness during disinfection.

Calcium hypochlorite is used considerably due to its high concentration of chlorine. Calcium hypochlorite products are sold at 65% active as opposed to the 10, 12, and 15% concentrations of the industrial grade sodium hypochlorite products. The higher concentration of the calcium product is somewhat misleading since the solubility of the product is severely reduced in hard water due to the high concentration of calcium already in the water. If calcium hypochlorite products are used they should be dissolved in water prior to addition to the well. Even then the solution will precipitate calcium carbonate in waters with high calcium and alkalinity. Since this very fine precipitate will lodge in pore spaces throughout the well and surrounding formation, use of excessive doses or regular additions can result in severe blockage of flow areas.

Chlorine is a strong oxidizing chemical and as such can be both effective at removing organics and a hindrance because of the by-products of such a reaction. When the hypochlorite ion is present the primary reaction is oxidation and the oxidative effect on organic material is varied. Chlorine by-products are notorious for being carcinogenic and the oxidation effects on polysaccharides are also a problem. The chlorine oxidation of a methane base chemical such as a hydrocarbon or humic acid is a trihalomethane or THM. The partial oxidation of a polysaccharide generally produces a gum material. This gum or oxidative by-product is discernable in the laboratory as a darkening or thickening of the upper layers of a biofilm and is thought to be the reason biofilms exert a certain resistance to chlorine (Le Chevallier and Babcock, 1987). It is also true that further oxidation of the biofilm degrades the organic material to its basic building block and as such is an excellent food source for bacteria (D.Vander Kooji, 1985; Applegate and Erkenbrecker, 1989). The bacterial regrowth often observed following chlorination is attributed to this abundance of food, however the inoculum for such an upsurge in growth comes from bacteria protected under the partially oxidized polysaccharide layer.

Chlorine application (see Chapter 1) is best carried out by preparing a concentration of chlorine at 50 to 200 mg/l concentration in a volume equal to four to five times the standing well volume. This supplies sufficient solution to flood the entire well environment. pH adjustment should be used to provide the hypochlorous ion (see Figure 9.11) which is considered at least 100 times more biocidal than hypochlorite (Pontius, 1990). To accomplish this adjust the pH of the water prior to addition of the sodium hypochlorite. The pH is usually taken to a pH of 4.5 to 5.0. The sodium hypochlorite, which is a strong alkaline pH, will raise the pH as it is added to the pH adjusted water. A rule of thumb is that the pH of the well water is usually raised between 1.5 and 2.0 pH levels for every 200 mg/l hypochlorite added. Most acids such as sulfamic, hydrochloric, phosphoric, and hydroxyacetic or glycolic can be used. Small well applications can utilize vinegar which is 5% acetic acid. Of the commercially available products those which are based on water pH and alkalinity measurements are most effective.

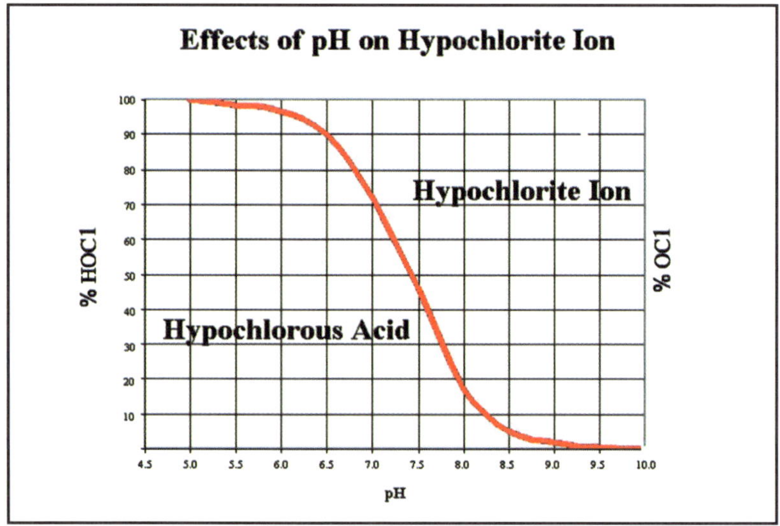

Figure 9.11. Effects of pH on hypochlorite ion

Products like Layne's QC-21, Johnson Screens' NuWell-310 and NuWell-410, and Design Water Technologies' ChloraPal also add dispersant and penetration additives which further enhance the disinfection properties. The following formulae can be useful in calculation of the pH control application:

Table 9.2. Calculations for pH control of chlorination

Vinegar: 1 gal. of vinegar will adjust 100 gals. of water where average alkalinity is 100 mg/l and the chlorine dose is 500 mg/l

				Desired Chlorine		
Gallons of vinegar	=	Alkalinity	X	levels	X	Volume
		100 mg/l		500 mg/l		100 gal.

Hydroxyacetic: (Glycolic acid) 1 qt. (or liter) will adjust 1,000 gals. of water with an average alkalinity of 100 mg/l and a chlorine dose of 200 mg/l

				Desired Chlorine		
Quarts of Hydroxyacetic	=	Alkalinity	X	levels	X	Volume
		100 mg/l		200 mg/l		1,000 gal.

WARNING: Chlorine gas is produced at a pH of 5 or lower. Care *must* be used when adding a hypochlorite solution to the water adjusted to a pH of 4.5 or 5.0. Blend only in a well ventilated area or outside. Stand up wind as the hypochlorite is added to the pH adjusted water. Neutralization above 5.0 will be immediate. Since chlorine gas is heavier than air this procedure must not be in a vault or low area.

Once the solution is prepared it is tremied into the well starting at the bottom and moving upward. Mechanical agitation at this point is then used to dissipate the chlorine throughout the well environment. Surge blocks or tight fitting swabs appear to be the most efficient, however many different applications are used particularly in smaller wells. Recirculation of the chemistry out of the well and back down hole is often used in smaller well systems. While this is useful in disinfection of the upper structure of the well, very little agitation reaches the deeper zones where coliform organisms most often cohabitate with anaerobic organisms. Since most pathogenic microorganisms would enter the deeper anaerobic biofilms any attempt at disinfection of a contaminated well must include surging at the lowest levels of the well.

Application of Rehabilitation Chemistry

A review of many well rehabilitation projects has shown that the most successful include preparations of the cleaning chemistry above ground. Initially, wire brush and evacuate the debris from the casing and screen areas. After addition of the chemistry by treming into place either through a separate line or directly through the surge block, the well is surged using a double surge block or tight fitting swab. The volume of chemistry prepared should equal from 1.5 to two times the standing well volume. This volume will supply sufficient chemistry to saturate the gravel pack and immediate formation. Use of this same calculated volume for open borehole constructed wells will provide a good level of chemistry to penetrate the formation around the well. Some cleanup of the surface of the borehole wall should be carried out prior to addition of the cleaning chemistry. This is usually best accomplished with the use of jetting tools with pressure adjustments made according to the stability of the surface. Once the surface has been cleaned any debris that has collected in the well bottom should be evacuated.

While chemical rehabilitation, decontamination, and disinfection are all often thought of as different cleaning methods for wells, they all should contain elements of initial scrapping and debris removal just as you would scrape a dinner plate before you put it into soapy water. Also, as you would swirl a dishcloth over the surface of the plate to ensure cleaning, you should mechanically work the well to apply the chemistry to the surfaces in the well. Pots and pans need soaking and most wells require application of the chemistry in the same manner with the chemistry left in the well over night and the mechanical action reapplied the next day before pump out.

CHAPTER 10 Concentration of Chemicals

A well contractor once asked me to review a couple of well rehabilitation failures. One well was in an obvious hard water area meaning it likely had a moderate to high level of calcium carbonate deposits in the gravel pack and screen sections of the well. The other was a job in which the well owner, a small municipality, was furious at the extended pumping required to return the pumped water to the normal pH.

A review of the chemicals specified and calculated for the individual projects revealed that the well in the hard water area had been treated with such a small volume of acid that the suspected carbonates most likely neutralized the acid in the first hour, if not sooner, and the subsequent two days of surging did little to remove the deposit. The second well was treated with 15% acid (based on well volume) and was in an area where the groundwater had a pH of less than 6.0.

Interestingly, the contractor who was new to both regions did not take a look at the water chemistry. Also, because of the bid specifications, the contractor used two different formulae to calculate the acid volume and concentration required. This resulted in considerably less chemistry used in the area of high hardness and considerably more being used in the area of no potential for carbonate precipitation.

Two factors are important in calculating the chemistry required for well rehabilitation in addition to selection of the chemicals required for the project. First, the volume of cleaning solution should be determined and, secondly, the concentration of the chemical to be used must be chosen based on the well water analysis. In Chapter 9 well parameters were defined in order to clarify the area that will have to be cleaned. In order to clean this area the volume will have to be calculated so that the cleaning solution volume can be properly determined. Solution volume is important not only because it is used to reach the areas to be cleaned, but also sufficient volume is necessary to solubilize the blockage and be able to carry it out of the well.

Determining the Total Treatment Volume (TTV)

Successful well cleaning depends on blending the chemicals above ground in sufficient volume to fill or flood the entire area to be cleaned. In extremely deep wells chemical volumes are often limited to a specific zone to be cleaned. In these situations careful calculations should be made to determine the exact area, designing a solution of sufficient strength and volume to flood that particular area and utilizing techniques to contain the solution in that zone. Strategically placed packers (see Figure 9.4) or the use of double surge blocks are examples of methods used to control the placement of cleaning fluids. In most wells the total well volume within the parameters to be cleaned is considered the total treatment volume (TTV). This calculation in the typical gravel pack well will take into consideration the standing well volume (SWV) and the volume in the gravel pack less the porosity factor. A shorter version which works well for average gravel packs less than six inches is to multiply the SWV by two for wells of six inch diameter or smaller and to multiply by 1.5 the SWV of wells greater than six inches in diameter. This calculation accounts for the water in the gravel pack and some of the surrounding formation.

Open borehole wells have less available pore spaces for mineral accumulations and it would seem have less need for extended volume for cleaning, but the need is in the carrying capacity of a larger volume of cleaning solution. In cases with open borehole construction use a factor of 1.3 to 1.5 times the SWV to calculate the required chemistry. In open borehole construction where occasionally well walls are heavily coated with both biological and mineral incrustation, two separate applications of chemistry are often more successful than a single application because more volume is necessary to solubilize the incrustation and remove it from the well. In this situation use half the chemistry in each application but use the same TTV for each application.

Each well is different. Some have extra large gravel packs constructed in a time when larger packs were thought to provide more filtration for the water. Some wells, particularly in Texas, have large underreamed areas near the bottom or water producing zones. These have extremely thick gravel packs and thus large areas of potential plugging and of consequence require more cleaning volume. Wells with gravel packs of various sizes larger than six inches should have the gravel pack volume adjusted for porosity then added to the SWV for a more accurate approximation of the TTV required. Some overestimation for penetration into the surrounding immediate formation is also useful. The objective in calculating the TTV is to prepare sufficient volume of cleaning solution to reach all the affected areas and to provide sufficient carrying capacity to remove the dissolved debris from the well.

Comparison of TTV Calculations

A well is 1,200 feet deep, 12 inches in diameter, with a static water level of 150 feet. The borehole diameter is 20 inches with a gravel pack of 30% porosity sand. The calculated volume necessary for chemical treatment is determined both by the 1.5 factor and the actual capacity figures and is shown in Tables 10.1 and 10.2.

Table 10.1. Total treatment volume (TTV) calculation

1,200 ft. - 150 ft. = 1,050 ft. of water
1,050 ft. x 5.9 gal./ ft. = 6,195 gallons
6,195 gal. X 1.5 = 9,293 gallons of cleaning solution

Table 10.2. Calculation of actual treatment volume

16.3 gal./ft. – 5.9 gal./ft. x .30 (porosity) = 3.12 gal./ft.
3.12 gal./ft. x 1,050 ft. = 3,276 gal. in gravel pack
5.9 gal./ft. x 1,050 ft. = 6,195 gal. SWV
SWV + gravel pack volume = 9,471 gallons

Determining the Concentrations of Chemicals

Just as proper calculation of the treatment volume is necessary, a review of the water analysis is required since it will indicate the type of potential mineral formation in the well. Knowing the potential for certain scale will dictate the concentration of acid to use. Mineral acids are by far the most efficient chemistry for removing mineral based deposits or at least the bulk of the deposits. They are the primary constituents in the cleaning chemistry formula. While stoichiometric calculations would allow us to calculate the exact amount of acid required if we know the amount and type of deposit, those parameters are almost impossible to determine. Over the years certain concentrations of mineral acids have proven more efficient when dealing with certain types of deposits. These concentrations are shown and discussed in Chapter 9 and repeated here for convenience.

Table 10.3. Acid concentrations based on water chemistry

Condition	Acid concentration
High carbonate and sulfate potential	10 to 12%
Some carbonate/sulfate or strong iron/manganese	8 to 10%
Moderate mineral potential with no heavy deposits	6 to 8%
No mineral deposit expected, water pH below 7.0	3 to 5%
No alkalinity and pH below 6.0	3%
Recent acid cleaned but using a biodispersant for biogrowth removal	1%

Examples of additives used to increase cleaning efficiency are the biodispersants and surfactants discussed earlier. Biodispersants, particularly the stronger products (NuWell-310 and NuWell-320, QC-21, and Catalyst), are generally used at a level of 0.5 to 3.0% of the total. The 0.5% level is usually restricted to rinse outs or cleaning of long distribution piping where continuous circulation is available. Levels of 1.0 to 3.0% are varied depending on the biofouling with the 3.0% level being used most often as it offers considerable improvement in the cleaning and removal of mineral solids as well as biofilm from the well.

Surfactants are used primarily to improve penetration and are used up to 0.5% of the TTV depending on the strength of the surfactant. Some dilute forms of surfactant products could require a higher concentration. They may be used in conjunction with the acid and the biodispersants but do not aid in biofilm removal with the exception of aiding the chemistry in penetration of the formation and mineral blockage. Table 10.4 is an example of the calculations for the actual amount of chemical required for well cleaning.

Table 10.4. Calculating chemical requirements for cleaning

Data

TTV = 9,300 gallons

Acid required = 8.0% of hydrochloric

Additive required = 3.0% NuWell-310

Hydrochloric acid available at 30% Active, 9.5 lbs./gal.

NuWell-310 is 100% Active, 10 lbs./gal.

Formulae

$$\text{Acid required} = \frac{\text{TTV x Wt. of Water x \% Acid Required}}{\text{Wt. of Acid x \% Active Acid}}$$

$$\text{HCl required} = \frac{9{,}300 \times 8.3 \times 8}{9.5 \times 30} = \text{2,166 gallons of 30\% acid}$$

$$\text{Additive required} = \frac{\text{TTV x Wt. of Water x \% Required}}{\text{Wt. of Product x \% Active}}$$

$$\text{NuWell-310 required} = \frac{9{,}300 \times 8.3 \times 3}{10 \times 100} = \text{232 gallons}$$

Variations in Calculating Total Treatment Volume (TTV)

Total treatment volumes, which fill the area to be cleaned and account for some intrusion into the immediate formation around the well, usually approximate 1.5 times the standing well volume and are usually more than sufficient for most cleaning and rehabilitation projects. Occasionally, though, the procedure calls for more cleaning solution such as in well floodings and in well systems with casings of six inches or less which require two times the standing well volume to produce a volume sufficient for cleaning. Well systems with large gravel packs usually require the calculation of the actual volume.

Chlorination

There are a number of references in this book concerning chlorination because it is a very important subject in that it accounts for disinfection and will be utilized as part of most decontamination projects. The recommended volume for proper chlorination is between three to five times the standing well volume. This quantity has proven necessary since the entire well structure is required to be washed with a solution of sufficient strength to control bacterial growth with a strong chlorine level remaining after the disinfection period.

A highly recommended procedure involves the blending of a hypochlorite solution in a tank above ground at a pH near 6.5 in sufficient volume to equal four times the standing well volume and then the placement of the entire volume into the well (Holben and Gaber, 2002).

Chemical Recirculation Cleaning

This method of cleaning is used for shallow wells and is particularly effective where heavy fouling has occurred such as often seen in recovery wells, environmental remediation wells, or wells that have set idle for extremely long periods. The method consists of the placement of a packer slightly below the first slots or screen openings and the positioning of a submersible pump below the packer near the bottom of the well. A chemical solution approximately three times the standing water volume is placed in a tank near the well. The chemical solution is allowed to gravity flow into the well and recirculated back into the tank through the submersible pump. The process requires only occasional attendance and can be circulated for extended periods of time. The drain on the tank should be placed sufficient distance off the bottom so that solids or settlings are not recirculated down hole.

A variation of this application is the simple recirculation of chemistry down hole. Without the packer there is no direct diversion of cleaning solution to the gravel pack; however application has proven very successful in wells heavily biofouled and/or hydrocarbon fouled, such as those in remediation systems or those wells with heavy iron oxidizer plugging of well screens. Again, these methods are most applicable in shallow wells (Mehmert, 1995).

Well Flooding

This method of cleaning has been successful when even vigorous surging has failed. It involves using sufficient cleaning solution volume to flood the well environment past the borehole wall in an attempt to reach the area of blockage. Usually the blockage at this point is composed of heavy biofilm development. The area is flooded with a cleaning chemistry, usually a blend of a mineral acid and a biodispersant, and allowed to set for several days. Surging prior to and followed by strong evacuation is used in the removal of the cleaner from the well when the cleaning program is complete. In order to have sufficient chemistry to reach areas of blockage, volumes of six to ten times the standing water volume are used (Mehmert, 1995).

CHAPTER 11 Mechanical Procedures Necessary for Complete Cleaning

I use the term "complete cleaning" as an application of chemistry to a well which includes pulling the pump, removing the debris and applying physical agitation. So many well owners and even some well professionals often try to clean or disinfect a well without pulling the pump and using mechanical agitation along with the proper chemistry. The injecting of chemistry downhole and bumping the pump, maybe even letting it set over night, will not provide complete cleaning and/or decontamination of a well. Even the use of mechanical application, such as the recirculation of a chlorine solution down hole, is not a completely effective chlorination or disinfection procedure. In very shallow wells where the pump is placed near the bottom, this procedure does result in some cleaning and/or decontamination, particularly where iron-oxidizing bacteria are the chief concern and their area of fouling is on the screen and casing surface. Most of the time the application of chlorine in this manner, however, does more harm than good. The oxidizing effect, especially on heavy biofouling, can result in less soluble formations which harbor coliforms or even pathogens very effectively.

As long as wells continue to be used we will always be tempted to add chlorine solution to a well in an attempt to remove an odor, taste, or offending organisms without first removing the pump and evacuating the debris in the well. Sometimes we will succeed, that is we will mask the organism's presence and temporarily solve the problem. Other times several attempts will at last be followed by a cleaning program involving removing the pump, evacuating the debris accumulated in the well, and then the systematic addition of chemistry and the application of mechanical agitations of some type. This is a fact not because it happens, but because you can't chemically clean something without first removing the build up of material and then applying the chemistry to the surfaces to be cleaned. The following are some of the mechanical applications or tools that are often necessary in the thorough cleaning of well systems.

Pull the Pump

This is an expensive procedure but becomes even more expensive as time is measured in incomplete cleaning and the resulting loss of the use of the well. Pulling the pump allows access to the well. If a video is warranted it is now much easier to perform. Pulling the pump also protects the pump from the harsh chemical that often must be used to achieve a good cleaning. Once removed the pump can be examined and if necessary repaired. Chemistry can be placed in the well where required and surge blocks and other necessary equipment can be used.

Physical Cleaning of the Casing and Screen

Wire or nylon brushing of the casing and screen will clean off heavy debris and allow it to be removed through bailing or airlift pumping. This debris which is a mixture of corrosion by-products, mineral deposits, and biological accumulation must be removed not only because it will place demand on the cleaning chemistry, but to allow the chemistry access to clean the surface of the well structure.

Evacuation along with brushing will remove corrosion cells, blockage by mineral deposits, and much of the bacteria accumulated in the well bottom. This particular cleaning phase is often the procedure responsible for removing those residual populations of bacteria which lead to quick resumption of the problem. Proper

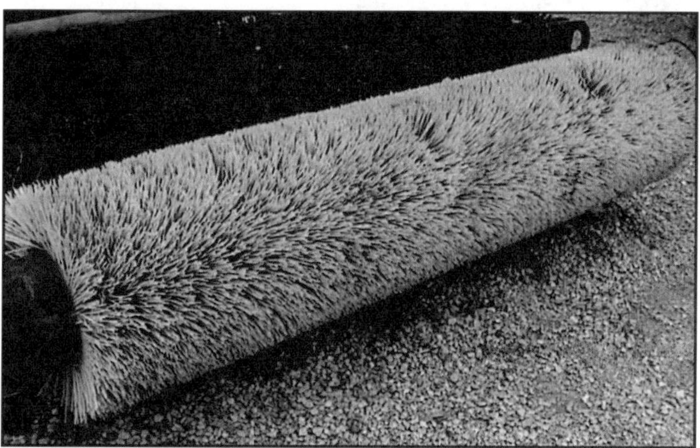

Figure 11.1. Typical nylon brush for large diameter wells

application and choice of chemical during the following phase will not only provide cleaning beyond the confines of the well proper, but will neutralize corrosion cells and provide passivation of the metal surface of the well structure slowing down future corrosion activity.

Figure 11.2. One type of wire brush

Surging

Webster defines surging as "a heavy violent or swelling motion." This motion is produced in a well through the use of a surge block or tight fitting swab that is raised and lowered in the well forcing the water to be pushed down the casing through the screens or slots into the gravel pack. The water is then pulled back when the tool is withdrawn up the casing.

Through the years this has proven to be one of the best physical forces devised for well cleaning. Laboratory studies of the water movement and corresponding cleaning with surging application of the chemistry show movement in the gravel pack and immediate formation. These are the primary zones of biological and mineral blockage. The volume movement of the chemical solution both dislodges and dissolves blockage in pore spaces and the directed force or flow moves this material back into the well proper to be removed. In addition to this mass movement of cleaning solution throughout the well environment, there is a very efficient localized cleaning of the casing and screen surface due to the up and down action of the tool.

Time spent on surging varies greatly between contractors and other well professionals as well as on the specific wells. It is good practice to standardize this operation. Time should also be spent on the bottom zone in sufficient time allotments to satisfactorily clean the surface and the sump or well bottom.

The Surge Block

The surge block is one of the best and most universal methods for applying mechanical activity during rehabilitation or chemical cleaning. However some guidelines must be exercised to insure good or even adequate results. Using a surge block implies a ridged tool with some pliable material incorporated within the construction to allow a relatively tight fit without subjecting the casing or screen surfaces to mechanical damage. The tighter the fit the more physical surface cleaning is achieved and the more force is applied to water movement both in and out through the screen or flow spaces. It is at this point however that damage can occur. If adequate rubber or other pliable material is not used damage will result as the metal structure of the surge block comes into contact with the walls of the screen. If the fit is too loose insufficient force will be applied during both the up and down strokes. On the other hand, if the fit is extremely tight and excess force is applied on the up stroke, collapse of the screen will occur.

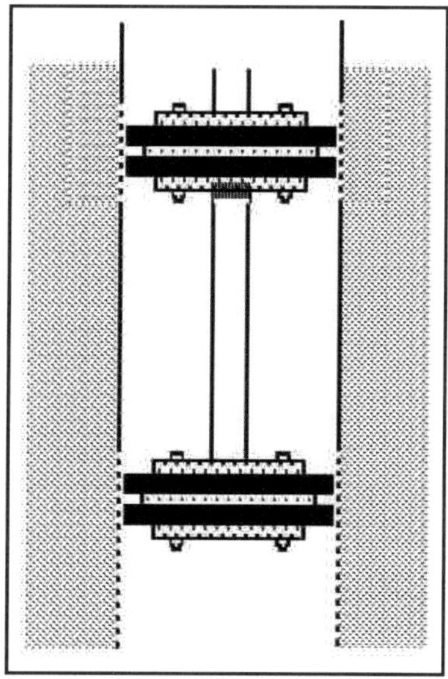

Figure 11.3. Drawing of a double disc surge block

The following are some guidelines to follow and I suggest you explore even more the mechanics of well surging as these suggestions are only a cursory guide.

Surge blocks are usually constructed in a double plate configuration (see Figure 11.3). They are constructed of steel; however, in the past wood was specified to limit damage to the walls of the casing and screen. The plates on each end are sized two inches less in diameter (many prefer one inch) than the diameter of the narrowest point in the well construction. A rubber gasket is sandwiched between plates at both ends of the surge block. The diameter of this gasket material is such that it is approximately one inch less than the well screen and/or casing diameter. (Here again some use the same size as the well screen diameter.) The unit should weigh a minimum of 150 pounds and have provisions to add additional weight if necessary in order to achieve adequate downward force.

While I have described the basic double surge block, there are many variations which often improve the work that can be done. A chemical or tremie line can be installed through the top plate which allows chemical injection into the zone between the plates increasing cleaning potential for that section as the surge block is moved up and down. A pump or suction line can be installed either above or below the surge block which allows a circular cleaning motion to occur during the upward and downward thrusts. Jetting devices have also been installed which allow pressure washing of the screen surfaces as well as some extra force applied to gravel pack cleaning. Not all additions improve the mechanical cleaning as some forces interfere with the water movements already begun by the surging effect.

The surge block should be operated slowly at first (0.5 to 1 ft./sec.) in order to blend the chemistry and place it into the gravel pack. The speed of the surging can be increased up to maximum of three feet per second to encourage more rigorous cleaning (Baird, 2002). The downward stroke must always match or exceed the upward stroke particularly in speed. If the upward stroke exceeds the force applied with the downward strokes (both in speed and length of stroke), movement of aquifer water toward the well will dominate resulting in fines and silts moving from great distances toward the well. This concentration of deposit in the circumjacent area of the well has considerable effect on the water flow available for well operation (Troutt, 2002).

Equipment Necessary for Good Surge Block Operation

When it comes to the mechanical side of chemical cleaning or rehabilitation of a water well the cable tool rig is without equal. The characteristic up and down drilling action of this type of drilling rig makes it ideal for surging actions. This action is imparted to the drill line (and in this case the surge block) by the walking beam. The walking beam pivots at one end while the other end is moved up and down by a single or double pitman connected to a crank-shaft. The vertical stroke of the walking beam, and thus the surge block, can be varied by adjusting the position of the pitman on the ball gear and the pitman connections to the walking beam. Changing the speed of the drive shaft can vary the number of strokes per minute. The control of both the speed (feet per second) and the vertical stroke (the area covered by each stroke) is critical in good surging technique (Driscoll, 1986).

Pump service rigs can also be used effectively for surging applications. They however are best if equipped with a "walking beam" which gives them cable tool capabilities and allows the speed of the beam (and the surge block) to be controlled by the speed of the engine. If a pump service rig without a walking beam or a rotary rig is used to surge a well, the speed of the surge block is controlled by the cable setup. Rigs with two part lines are one-half the speed of the single line and those with more have their speed reduced by half for each additional line. Rigs with hydraulic drives are also controlled by the number of lines used, thus a three part line would have the speed reduced to such an extent that surging would be unsuccessful. Even if the rig has a straight or one part line the speed is limited by the hydraulic control system. The maximum speed developed may be inadequate for surging applications.

The chemical cleaning operation requires the input of energy from the operation of the surge block. This energy must be imparted over a certain period to successfully clean any given surface. In the cleaning of either screens or casing the time working a certain area is dictated by the degree of fouling both on the surface (which is left after wire brushing) and behind the screen in the gravel pack and formation. The time interval is usually between one and three minutes per foot of screen and one-half to one minute per foot of casing. Surging is usually applied at this rate the first day and at one-half the rate after sitting overnight. Gravel packs or formation heavily impacted with gypsum or calcium sulfate may require a third day of surging.

Jetting

Jetting is the high pressure spraying of cleaning solution, water, or gas-water mixture through an apparatus designed to direct the spray at the casing and screen walls and through the screen/slot openings into the gravel pack. Since the pressure can be adjusted to deliver the necessary chemistry, cleaning of the immediate surfaces is excellent. Where wire wrapped screen is the screening material good penetration into the gravel pack is also accomplished. Shuttered or louvered screens often cause a deflection of the jetting force which reduces effectiveness in these applications (Driscoll, 1986). There are some modified units which are designed to jet downward at an angle allowing some force to enter the gravel pack. While this is an improvement, rehabilitation in wire wrapped screen wells is usually considerably more successful.

Jetting is used in all types of wells, however we have found it the method of choice in open borehole well construction. Use of surge blocks or swabs can damage the borehole wall if vigorous action is used to clean stubborn incrustations. Jetting can be controlled and used very efficiently to clean the surface and access areas of the open bore design. While the many different contractors have almost as many different techniques for jetting, the usual pressure used is between 100 and 300 psi with the spray nozzle positioned one-half to one inch from the borehole face. In some areas of the country however, pressures as high as 1,000 psi or greater are used depending on application technique and formation structure.

Figure 11.4. Jetting Tool

Nitrogen gas and carbon dioxide are often used with cleaning chemicals as well as with water alone to help dislodge stubborn deposits and to limit the volume of solution required.

Air, however, is not recommended in jetting applications since it will release air into the aquifer resulting in some physical blockage of flow areas as well as encouraging aerobic bacterial growth. Carbon dioxide could be a problem in areas with high hardness water since the addition of carbon dioxide to water results in the formation of the bicarbonate ion HCO_3^-. Bicarbonates react with calcium hardness to form calcium carbonate precipitate (Sienko and Plane, 1957; Manahan, 1994).

Re-circulation of the chemical solution during jetting is often used where provisions are made to filter or clean up the solution before circulating back down hole. Great improvement of overall cleaning is noticed if the cleaning solution is continuously checked both for solids accumulation and pH or chemical concentration. While in-line filtration is usually required, it should be first pumped to a tank for sedimentation or heavier solids removal prior to the filter. Where strong evacuation or pump out is used to draw extra water from the formation, some chemical will have to be continually or periodically added to maintain a strong chemical solution. *Safety should also be a concern when recirculating strong acid or caustic chemistry.*

Air Displacement Surging or Greiner Method

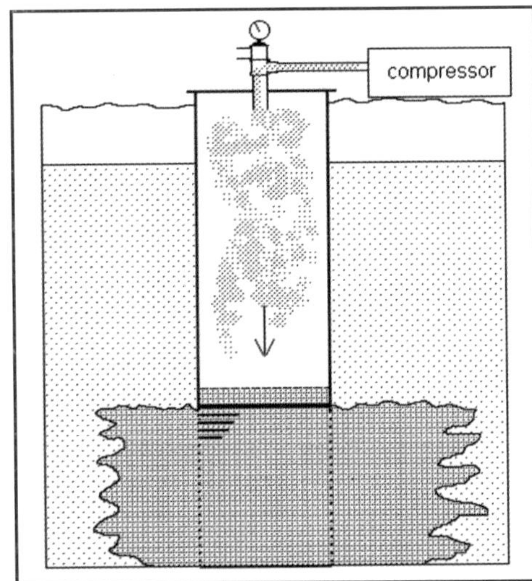

Figure 11.5. Greiner Method
Outlet valve is closed, well is charged with air to drive static level down to within 5–10 feet of the top of the screen. Pressure required (psi) is found by dividing the distance to the top of the screen minus the SWL by 2.31.

Air can be used as a means of surging the cleaning chemistry if the area between the static level and the top of the screen is sufficient to allow a large enough volume of water to be forced through the screen and gravel pack. The well should also be sufficiently shallow to allow even distribution of the displaced solution through the impacted screens. Figures 11.5 and 11.6 show the adaptation necessary for use of this method. A rule of thumb for this application: the static water volume in the casing above the screen should be equal to or greater than 1.5 times the screen volume. The length of the screen zone should be less than 60 feet. The following formula can be used to determine the air pressure required: depth of the water column above the screen divided by 2.31 = psi required to push the water to the top of the screen.

After placing the cleaning chemistry into the well an air tight seal is placed on the wellhead. A pressure gauge and air compressor are attached and air is pumped into the well forcing the cleaning chemistry downward and out through the screen section. Care should be taken not to force the cleaning solution below the top of the screen. The air is released through a large valve positioned through the seal on the wellhead and the process is repeated. Depending on the well size, the operation is usually repeated five to six times per hour and where used successfully was usually operated for two to four days. Be sure to check the chemistry to maintain the cleaner at a pH of 3.0 or below. Figures 11.5 and 11.6 depict the Greiner Method.

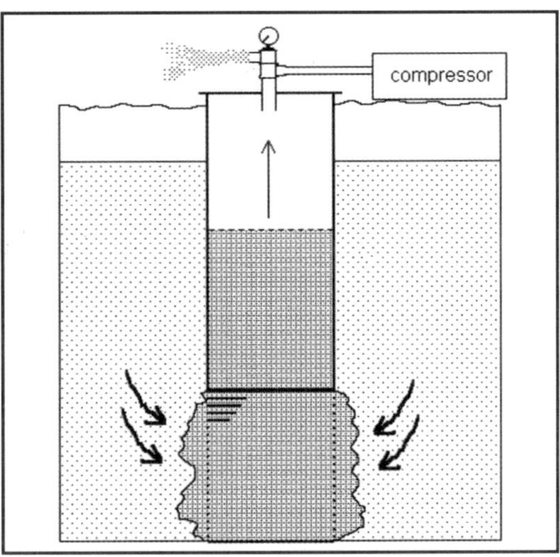

Figure 11.6. Greiner Method
Release of pressure allows fluid to rush back into the well. Result is excellent agitation in the effected zones.

Cyclic Flooding With Pump in Place

While flooding of the well system was mentioned in Chapter 10, a variation is worth noting that may be used on smaller wells without pump removal. The cleaning solution is prepared and maintained in a tank near the well. The cleaner is allowed to gravity-flow into the well through the pump. After a suitable volume is placed in the well, the pump is operated to return the solution to the tank and the cycle is repeated. The system can be placed on timers and left operating over an extended period which aids considerably in this type of a more passive cleaning program.

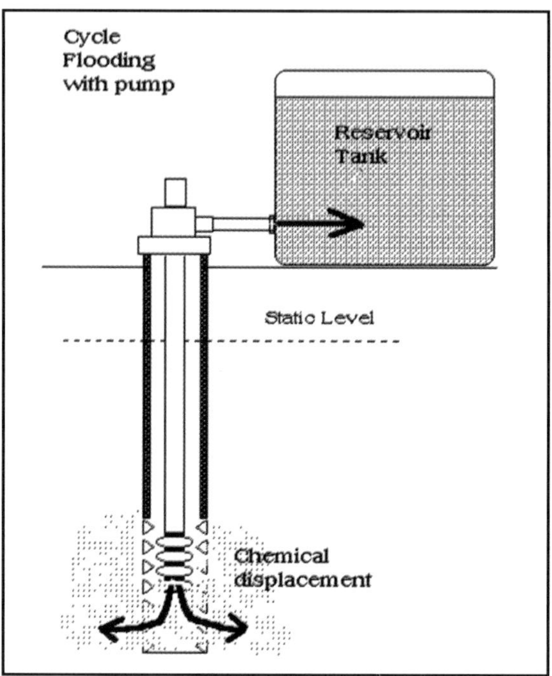

Figure 11.7. Cycle flooding through pump
The chemicals are allowed to flood the well by gravity flow through the pump. The pump is activated to return the chemistry to the surface tank.

At least three well volumes of chemistry will have to be used to assure adequate flooding of the system (see Figure 11.7).

Evacuation or Development

To a chemist evacuation is the pump out of a well, the removal of loosened or accumulated debris, spent chemicals, and/or any silts or sands that may have infiltrated the well. To a well professional the pump out or evacuation of a well is the mechanical manifestation of an all-important task of well development. "Well development," as once told to me by an old well driller, "is required to return the formation to its natural placement to repair the damage that was done while the well was being drilled." *Groundwater and Wells* (Driscoll, 1986) states that well development has two broad objectives: 1) repair damage done to the formation, and 2) alter the basic physical characteristics of the aquifer near the borehole so that water will flow more freely to the well. Most of the discussions about development are aimed at the well construction phase, but when a well has been chemically treated it is in need of much the same attention.

Changes have taken place in the well which have brought about the need for disinfection or chemical cleaning. Either heavy bio growth, mineral plugging or both has taken place in the well and zones around the well. The application of chemistry has not only removed some or all of the blockage, but has resulted in displaced debris accumulation and disturbance of the natural formation. The cleaning or chlorinating chemistry must be removed, but in addition sufficient effort must be made to remove all of the debris and return the formation to as uniform and open state as possible. Pump out of the cleaning chemistry must include sufficient energy to do this. In other words, sufficient pump out or development must be performed. Just as in initial development, the final objective must be to return the formation to the most open porosity and restore hydraulic conductivity.

In the past several years there has been a concerted effort to encourage "well development" following cleaning or before chlorination. This has been one of the single most effective steps which increased the success rate against coliform positive wells. We believe that the extra time spent on development is removing much of the debris and improving water flow to the extent that it limits bacterial accumulation in the well.

CHAPTER 12 Neutralizing Cleaning Chemistry for Discharge

As with any cleaning operation, a well rehabilitation project is not complete until all the cleaning chemistry is accounted for, properly neutralized, and discharged according to the local, state, and federal regulations. In well cleaning there are three areas of concern:

1) Neutralization of discharged acids or alkaline cleaners

2) Removal of suspended solids and particulates

3) Dechlorination of the disinfecting chlorine solution

All of these operations must be carried out on the surface once the cleaning or chlorine solution is pumped from the well.

Acid Neutralization

If an adequate concentration of acid was used and a pH of 3.0 or below had been maintained during the cleaning procedure, then the discharge from the well could be at a pH of 3.0 or below. This level of acid will require neutralization before discharge to a municipal sewer system, other approved discharge point, or hauled off to an approved waste site. In addition to meeting requirements, neutralization should also reduce fees for discharge or delivery to a hazardous waste disposal operation.

Table 12.1 shows several products which can be used to neutralize the discharged acid. The well pump out should be collected in a baker tank or some other suitable vessel in which the neutralization reaction can be carried out. Past use of a trough filled with lime is much too crude for today's regulations. Such operation often produced discharge that was above the allowable pH (excessive solubilizing of lime easily produces pHs above 10). Both lime and soda ash neutralization resulted in excessive solids discharge but much of these operations were not monitored closely, particularly for dissolved and suspended solids.

Table 12.1. Acid pH neutralization chart

Approximate Quantities of Soda Ash or Magnesium Hydroxide to Neutralize Well Cleaning Discharge at Various Recorded pHs	
pH of discharge well water	Pounds of neutralizer required for each pound of acid used as cleaner
6	0.10 lbs. soda ash or 0.08 lbs. magnesium hydroxide
5	0.25 lbs. soda ash or 0.15 lbs. magnesium hydroxide
4	0.50 lbs. soda ash or 0.40 lbs. magnesium hydroxide
3	0.75 lbs. soda ash or 0.60 lbs. magnesium hydroxide
2	1.00 lbs. soda ash or 0.80 lbs. magnesium hydroxide
Soda Ash – (Na_2CO_3) – will release carbon dioxide during the neutralization so some frothing or reactivity will take place; avoid tight spaces	
Magnesium hydroxide slurry – $Mg(OH)_2$ – 55 gal. drum is hard to handle but product is excellent for neutralizing; minimum heat and danger to personnel.	
Lime – ($Ca(OH)_2$ or CaO) – not very soluble; will cause some heat if acid pH is very low; dust can be dangerous for eye contact; usually considerable fines or insoluble particulate remains.	

The vessel chosen for the neutralization operation should be large enough to allow some control so adjustments to the neutralizing chemistry can be made to assure the discharge is within the accepted parameters. If discharge is going to be made direct from the neutralization tank then the discharge outlet should be placed some distance off the bottom of the tank so space remains for the collection of any settled solids. These can later be pumped off and hauled to a suitable landfill or other acceptable disposal site. Some contractors who have dedicated tanks for neutralization place baffles in the tank to prevent short-circuiting of the mixing. This provides a much more even and more easily adjusted neutralization. Baffles also aid the sedimentation and flocculation process required where solids must be controlled.

Lime and soda ash have been the neutralizing chemicals of choice as they are inexpensive and readily available. Lime adds weight to the settling floc used to settle out suspended solids and soda ash has a natural buffering effect. This buffering prevents an excessive rise in pH due to over treatment. As the process for controlling discharge becomes more sophisticated, the liquid alkaline products such as sodium hydroxide (liquid caustic) and potassium hydroxide (caustic potash) as well as magnesium hydroxide are being specified and used more often. The liquidity of the first two makes them easy to pump into blending tanks and require no preparation. The magnesium hydroxide is a 50 to 60% slurry and has the added advantages of reducing the discharge TDS level and providing a natural buffer at a pH of 9.0. As the pH is raised toward neutral or the accepted pH range for discharge, most acid cleaning solutions will precipitate at least 50% of the solids (dissolved downhole) and the use of either lime or soda ash will increase the volume. While sodium and potassium hydroxide have the advantage of ease of use producing primarily only soluble products of neutralization, the TDS usually increases with the soluble sodium and potassium products.

Calculation of the neutralizing chemistry is not as easy as it should be; because while the pH of the discharge water can be determined, the amount of water to be discharged is usually unknown. There is also considerable neutralization of acid down hole and calculation of the remaining acid remains a mystery. Table 12.1 is a chart showing the potential neutralization requirements based on the amount of acid used and the pH of the discharge water. Keep in mind, though, that the water pH will be changing as the water is pumped from the well and so the chemistry will have to be added as needed to reach the necessary pH point. The values shown were determined by empirical (that is to say, often repeated simulations in the lab) testing and should be used as an estimate providing an edge to the high side. The chart should be very useful in determining the quantity of neutralization chemistry to have available or to bid for a specific job.

The actual process will require the use of a pH meter or pH paper or test kit in order to adjust the amount of neutralization product so that the discharge to the sewer, etc. is maintained within acceptable

Figure 12.1. Acid neutralization tank with adjustable baffles Soda ash is added by means of a hopper (not shown).
(Reprinted by permission of Water Well Renovations, Co.)

limits. Some contractors prefer to neutralize in batches and others use a continuous operation which works well, particularly if no solids removal is necessary. Where solid removal is required more time may be required for the sedimentation or flocculation reaction to be carried out. If multiple tanks are used it is usually easier to set up a continuous operation. Figures 12.1 and 12.2 are examples of different types of tanks used for neutralization.

Neutralization of Alkaline Discharge

Usually alkaline cleaning of a well system is used to remove oil or hydrocarbon-based contaminants. Discharge may be restricted even with pH neutralization, however the pH neutralization should provide some reduction in fees charged at the disposal facility. If sodium or potassium hydroxide is used as part of the cleaning chemistry, most likely the concentration used was between 3 and

Figure 12.2. An acid neutralization tank with auxiliary pump Fluids can be recirculated for better mixing.

5%. Since very little neutralization of the pH would have occurred down hole the requirement for acid is relatively easy to compute. You should choose an acid for the neutralization process based on the premise of producing as little precipitation as possible. Precipitation translates into solids accumulations which would be another factor to deal with as a possible disposal item. While hydrochloric acid is an excellent choice (since minimal precipitation would be expected), often times sulfamic acid and sodium bisulfate are used particularly on small jobs due to their dry powder or crystal form and ease of transportation and handling. The quantity of hydrochloric acid necessary is relatively easy to determine as approximately one part of hydrochloric 100% active would neutralize one part of 100% sodium hydroxide and 1.5 parts 100% potassium hydroxide.

Table 12.2. Caustic neutralization

Examples:
If 500 lbs. of 50% NaOH was used: 500 lbs. x 50% = 250 lbs. of 100% NaOH Neutralization = 1:1 250 lbs. NaOH (100%) = 250 lbs. HCl (100%) If 30% HCl is available, then: $250 \text{ lbs.} \times \frac{100\%}{30\%} = 833$ pounds of 30% HCl
If 500 lbs. of 45% KOH was used: Neutralization = 1.5:1 500 lbs. x 45% = 225 lbs. ÷ 1.5 = 150 lbs. of 100% HCl If 30% HCl is available, then, $150 \text{ lbs.} \times \frac{100\%}{30\%} = 500$ pounds of 30% HCl

NOTE: If a dry product is preferred, use sulfamic acid or sodium bisulfate according to the following ratios: one part potassium hydroxide (45%) will require one part sulfamic acid or one part sodium bisulfate. One part sodium hydroxide (50%) will require 1.25 parts sulfamic acid or 1.5 parts of sodium bisulfate.

CAUTION: Powdered chemical should always be dissolved first before adding the chemical to the solution to be neutralized. Most neutralization is exothermic, that is, heat will be produced.

Physical Operation

The actual neutralization should never be performed down hole as this will promote considerable precipitation or fallout. The caustic discharge from the well is neutralized in much the same manner as acid neutralization is performed. A baker tank or other vessel should be used to collect the pump out and the dissolved or liquid acid can be metered into the flow allowing neutralization to take place as the well is being pumped. The neutralized chemistry can be monitored before discharge to a sewer or other acceptable receiving source. A second tank can facilitate monitoring and collection of settleable solids if required. Again, if discharge is to be made direct from the blending tank, then the discharge valve should be some distance from the bottom to allow collection of the solids and use of baffles will prevent short-circuiting of the neutralizing chemical.

Solids Removal

More and more state and local if not federal regulations are being made to control the discharge of suspended or settleable solids. Gradually these regulations are being adapted to the discharge of any water and this includes the discharge following a well cleaning or rehabilitation project. The very nature of cleaning is the removal of solids and those solids are being moved out of the well and discharged. Many of the solids are completely dissolved but others are suspended and will remain suspended as they are discharged to a receiving water. This will add considerable load to a waste treatment plant or contaminate a receiving stream or waterway. In addition the neutralization of the pump out will result in many of the dissolved products either precipitating or remaining in the water as suspended solids. Fortunately, the neutralization process and the precipitation of solids can be the first step in solids removal. As solids are formed they reach a density sufficient to cause them to fall out of solution. This process can be encouraged with the use of certain flocculating agents that cause the particle to pull together becoming larger and more dense. The heavy particulates then fall out of solution or settle clearing the solution to be discharged. If the reaction is handled correctly the resulting floc will capture any organic matter (bacterial debris often suspended) and clay materials removing them from the water. Flocculants are classified as anionic (negative charged), cationic (positive charged), or nonionic (neutral charged). Usually the anionics and the nonionics are the most effective but the use of polymer flocculating agents is more of an art than science and therefore several should be tried until an effective one is found. You should also have others in reserve as different discharges may require different flocculants.

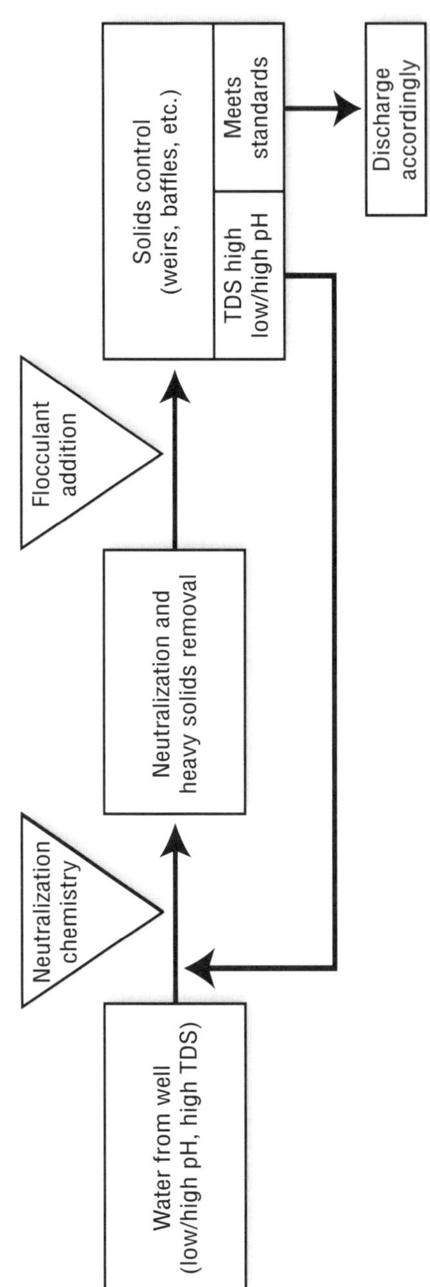

Figure 12.3. Schematic of discharge treatment
pH neutralization and solids removal treatment of well cleaning discharge water

Flocculants need to be added in fairly dilute solutions and at a point where mixing takes place so that they come in direct contact with the solids to be settled. As Figure 12.3 shows, the flocculant should be added after the neutralization chemistry is added. This is important because the neutralization will increase the fallout of solids and is necessary to be acted on by the flocculant causing the increase in size which encourages the settling process.

Mixing is the next step in importance. In systems using one tank for both neutralization and flocculation, you should observe the flow to see if mixing takes place after the flocculant has been added. If not, you can improve mixing by adding a mixer or by discharging into the tank through a pipe constructed with an in-line mixer. The neutralizer is added in the first part of the pipe and the flocculant is added several feet downstream. This method is advantageous as no additional power requirement is necessary for a mixer, however neutralization must be controlled closely so that maximum solids are precipitated without an excess pH rise. In the configuration where two tanks are used the flocculant can be added as the neutralized water is allowed to spill over into the second tank. The passing of the neutralized water to the second tank and the mixing of the flocculant can be facilitated by passing the water through an in-line mixer, as mentioned earlier, which should provide the contact necessary to start the flocculation process. The process can be further tuned by directing the pipe discharge to cause the water in the second tank to swirl in a clockwise direction (works well for round tanks). This will encourage flocculation and settling.

Water should be discharged from the tank at a point no lower than one-third the distance off the bottom, however a series of discharge ports can be installed from 12 inches off the bottom upward to facilitate discharge above the solids blanket. As the sludge is formed a blanket or mass develops and moves downward. The sludge or settleable solids closest to the bottom are the most dense, therefore the top part of the blanket contains the lighter material still growing in size and density. You should position your discharge so that this material is not siphoned off as you discharge from the tank. Hopefully you will be able to operate on a continuous basis, but you can operate in batches stopping the discharge into the tank and allowing the solids to completely drop out before you discharge from the tank (see Figure 12.3).

Chlorine Neutralization

For years the discharge of high concentrations of chlorine from well disinfection projects have gone unquestioned. Regulatory activities, however, are catching up and now the discharge of very low levels (as little as 1 mg/l) can result in a fine. This certainly is understandable since highly oxidative chemistry, especially chlorine, has been implicated in the formation of carcinogenic chemicals. The expulsion of high levels into the environment could result in the production of many unwanted by-products.

A form of reducing chemistry is necessary to neutralize the highly oxidative hypochlorite solutions. The chart marked Table 12.3 shows a list of suitable products that can be used to carry out chlorine neutralization. The formula in Table 12.4 is relatively easy to use for a batch application. Continuous neutralization of the chlorine while pumping to waste may take some additional calculations and preparations.

Table 12.3. List of dechlorinating agents

Theoretical Amount of Dechlorinating Agent Required for the Removal of One Part Per Million (mg/l Chlorine)			
		mg/l required	**lbs. per 1,000 gals.**
Ascorbic Acid	$C_6H_8O_6$	1.92	.016
Sulfur Dioxide	SO_2 (Gas)	0.90	.008
Sodium Bisulfite	$NaHSO_3$ (Powder)	1.46	.012
Sodium Sulfite	NA_2SO_3 (Powder)	1.77	.015
Sodium Thiosulfate	$Na_2S_2O_3$ (Powder)	0.70	.006

Example: 50 mg/l chlorine to be removed would require .012 lbs. x 50 = 0.6 lbs. of sodium bisulfite per 1,000 gals. of water to be treated.

Ascorbic acid is Vitamin C and can be purchased in food grade quality.
Sodium bisulfite and sodium sulfite are most often used for one-time dechlorinating projects because of availability and cost. The sulfite is usually the one preferred because of less disagreeable odor. For practical purposes sodium metabisulfite is the same as commercially available sodium bisulfite.

Reference: *Water Quality and Treatment,* AWWA, 3rd Edition, McGraw Hill 1971

In order to control the neutralization process pump the discharge into a tank. You can check the level of chlorine in the well water and calculate the neutralizer required once the discharge tank is full. The chlorine solution in the hole will decrease as the well is pumped due to the incoming aquifer water. This reduction in chlorine levels must be followed by a reduction in the neutralizing chemistry. While this is an excellent problem for calculus, I will attempt a more practical solution (see Table 12.4).

Table 12.4. Dechlorination problem

Problem:

The total treatment volume of the well (4 x the standing well volume)
= 1,600 gallons
The treatment level was 200 mg/l of hypochlorite
The residual or remaining chlorine is 160 mg/l
Therefore:
Let's assume the first 1,600 gals. from the well will contain
a chlorine level of 160 mg/l
Sodium sulfite is required at the rate of .015 lbs to remove 1 mg/l
chlorine per 1,000 gals. of water.

Formulas:

Rate, x the level of chlorine, x the volume, divided by 1,000
.015 lbs. x 160 mg/l x 1,600 gals./1,000 gals.
= 3.84 lbs. of sodium sulfite in the first 1,600 gallons of discharge.

Assuming you will pump from the well at the rate of 100 gallons a minute, it will take 16 minutes to pump the total disinfectant volume into the tank. Therefore the 3.84 pounds should be diluted to a quantity easily pumped into the water being discharged into the tank. For instance if you prepared 32 gallons of solution, you would pump two gallons a minute into the discharge stream as it entered the tank. In practice you should prepare 40 gallons of solution by mixing 4.2 pounds of sodium sulfite into 40 gallons of water and applying it evenly as you discharge into the receiving tank. At this point you should check the chlorine level both at the well and at the discharge point from the tank. If the chlorine has been neutralized in the tank then the water is ready to be discharged. If the chlorine level in the well has dropped to a percentage of the first reading (for instance a reading of 40 mg/l is 25% of the initial 160 mg/l) then approximately one pound of sodium sulfite can be prepared for the next batch. This figure should be adjusted if additional sulfite was used because some residual chlorine remained in the first batch.

We have essentially calculated two batch mixes; but by observing the chlorine reduction that you usually experience in the first pump out and the second and third you can precalibrate your batches of neutralizing chemistry and be ready to switch as you begin to discharge completely neutralized water from the receiving tank.

Some contractors may wish to use the batch method entirely as this gives them a chance to check each batch before they release it to the discharge point. It is good to allow an extra 10% neutralizing chemistry since the calculations on the chart are theoretical; however do not get into the habit of using excessive amounts of chemistry to "be sure" as there could be restrictions on their discharge at some point in the future.

Testing for chlorine residual is easily accomplished with the use of tests kits designed for large and small chlorine dosages.

Test Strips	Approximate Cost
HACH, 0 – 10 mg, #27450-50	$10.00
VWR, 0 – 120 mg, #EM-10043-1	75.00
Drop Test Kits	
HACH, 10 – 200 mg, #24444-000	$28.00
HACH, 0.2 – 4 mg & 1 – 20, #2254-01	55.00

It is imperative that you have some means of testing for chlorine both before and after the neutralization reaction.

CHAPTER 13 A Chemical Specification

This chapter is designed as a quick resource for a chemical rehabilitation project as well as a review of all the information given in the previous chapters. Where the project is a decontamination cleaning, then it is understood that the necessary work has already been done to establish the identity of the contaminant and the correct cleaning chemistry has been determined. For example: if the well had been contaminated with oil, then possibly an alkaline cleaning would have preceded the typical rehabilitation using an acid cleaning followed by disinfection.

Let's walk through this step-by-step.

Well Specifications

Well depth: 600 feet
Materials of Construction: carbon steel casing with SS screen
Well Diameter: 12 inches Static water level:125 feet
Screened from 250 feet to 550 feet Borehole: 20 inches

The water has high hardness, 268 mg/l, and an iron level of 2.1 mg/l. The alkalinity is 205 mg/l. Bacterial analysis was high and iron-oxidizing bacteria were noted on the microscopic evaluation. Since the well had been steadily loosing capacity over the past few years, it was decided to do a chemical rehabilitation along with a complete reworking of the pump and inspection of the drop pipe, casing, and screen.

Table 13.1. Chemical selection for cleaning program

Phosphoric acid (75%) at a 12% concentration in well
NuWell-310 Biodispersant 3% concentration in well
Sodium hypochlorite 200 mg/l (.02%)
NuWell-410 pH control chemistry for chlorination
Soda ash for acid neutralization
Sodium sulfite for chlorine neutralization

Table 13.2. Chemical calculation for cleaning program

12 inch well = 5.88 gals/ft 20 inch borehole = 16.33 gal/ft

SWV = 600 – 125 x 5.88 = 2,793 gallons

TTV (total treatment volume) = 2,793 x 1.5 = 4,190 gallons
x 8.3 = 34,773 lbs

TDV (total disinfectant volume) = 2,793 x 4 = 11,172 gallons
x 8.3 = 92,728 lbs

Phosphoric Acid:

34,773 lbs of water x 12% = 4,173 lbs x 100/75 = 5,564 pounds
of 75% Phosphoric acid

5,564 lbs/13.1 lbs/gals = 425 gallons of 75% Phosphoric acid

NuWell-310 (Biodispersant)**:**

34,773 lbs of water x 3% = 1,043 lbs of NW-310 or 105 gallons

Soda Ash

From the chart (Figure 9.8) 0.75 lbs x 4,173 lbs of acid
= approx 3,130 pounds of soda ash

Sodium Hypochlorite:

92,728 lbs x .02% = 18.5 lbs x (100%/10%) = 185 pounds of
10% sodium hypochlorite
185 lbs/8.5 lbs per gallons = 22 gallons of 10% sodium hypochlorite

NuWell-410 (Chlorine Enhancer):

Use at the rate of 1 gal per 1,000 gals of water for every 100 mg/l
of alkalinity present, and for every 200 mg/l of chlorine to be used
10,000 gals/1,000 gals x 205/100 = 20.5 gals of NuWell-410
(if 200 mg/l chlorine is used and water alkalinity is 205 mg/l)

Sodium Sulfite (See Figure 12.5)

.015 lbs/ 1,000 gals for every 1 mg/l of chlorine

1st batch (10,000 gals) will require approximately 20.5 lbs sodium sulfite estimating 33% loss of chlorine concentration during the disinfection process.

2nd batch will require 1/4 to 1/2 the amount required for batch one or 5 to 10 lbs

Estimate of total requirement = 35 pounds

Application of Cleaning Chemistry

On site a 5,000-gallon baker tank should be filled with approximately 3,800 gallons of water and to this add the phosphoric acid and blend. Add the NuWell-310 to the acid mixture and blend. Once the pump has been pulled and all necessary inspections performed, the well casing and screen area should be brushed to remove incrustations. The debris should then be removed from the well using an airlift or bailer. Be sure to remove all the debris from the well bottom. Tremie the acid dispersant mixture into the well placing approximately 20% (800 gallons) at the static water level, 40% (1,600 gallons) at the top of the screens, 20% at the half-way mark of the screen area, and 20% near the bottom.

Using a double disc surge block or tight fitting swab begin to surge the well starting at the bottom and working upward. Surging in twenty-foot segments spend at least 20 minutes per segment in casing zones and 40 minutes in screened areas. Check the pH at least once in the first hour and hourly until the pH stabilizes. If the pH rises above 3.0 then additional acid is required. Secondary additions of acid are usually made in 20% (of original volume) increments and can be placed directly at the top of the screen. Once the pH remains below 3.0 then the well should be allowed to sit overnight. The following day the surging should be repeated but only at one-half the time spent the first day. Evacuate the well as soon as the surging has been completed, do not allow to stand overnight. Begin evacuating near the static water level moving downward to the bottom, pumping each level until the water is clear. The well should be evacuated until the pH of the discharge is that of the original aquifer sample.

Neutralization of Cleaning Chemistry

All cleaning chemistry should be neutralized above ground, that is outside the well. The usual procedure is to pump the discharge directly into a baker or other storage tank in which the neutralization can be carried out. Since approximately 75% of the acid will be removed from the well in the first ten well volumes, much higher levels of neutralizing chemistry will be required at this time.

Proper discharge from the neutralizing tank will require maintaining a pH level between 5.0 and 9.0 (this should be checked as each locality often has its own regulatory level). You should have on hand a suitable pH meter or pH indicator paper to check your neutralization and assure proper discharge. Generally if discharge is being made directly from the neutralization tank, the point of discharge should be at least 12 inches from the bottom to insure a space for solids accumulation and prevent carryover of solids to the receiving system or body of water. You may wish to observe the level of solids your methods of rehabilitation and neutralization produces and vary the placement of the discharge pipe accordingly.

Application of Disinfectant Chemistry

The disinfectant volume can be approximated at 10,000 gallons (this will constitute 3.5 x the SWV) so that if a 5,000-gallon baker tank was used for the cleaning solutions it can be used to produce two batches of hypochlorite solution for disinfection. To the baker tank add approximately 4,900 gallons of water. To this add five gallons of NuWell-410 for every 100 mg/l of alkalinity in the water. The tank of water which has an alkalinity of 205 mg/l will require 10.25 gallons of the NuWell-410. After blending the NuWell-410 into solution, carefully add ten gallons of 10% sodium hypochlorite to the mix.

CAUTION: The addition of hypochlorite solution to a pH below 5.0 can produce chlorine gas. The procedure should be performed in the open. Stand with the wind blowing away from you when adding the chlorine solution. Neutralization above 5.0 will take place almost immediately.

Tremie the chlorine solution into the well at essentially three points: the upper casing above the static water level, the area between the static and the bottom, and the bottom of the well. You should start by adding approximately 20% to the top of the well washing any exposed structure, then add approximately 60% as you move the tremie line downward, finishing by pumping the remaining 20% into or near the well bottom.

Once the chlorine solution has been added, the solution should be surged or swabbed into place. Spend at least one minute per foot in the screened, louvered or slotted zones with less time (1/2 minute per foot) being spent on the casing. Let stand overnight and repeat the surging the second day at one-half the time spent the first day. Begin to airlift out of the well as soon as the surging is completed and the equipment removed from the well. Start at the static water level and work downward. Pump each section until clear or the chlorine is relatively low. Pump out should be continued until the chlorine level is 1 mg/l or lower. Be sure to evacuate the well bottom completely.

Neutralization of Chlorine Discharge

The pumped chlorine solution should be directed to a baker tank or other suitable storage tank where it can be neutralized or dechlorinated prior to discharge from the site. The sodium sulfite solution should be dissolved in a separate tank at the rate of one pound per gallon of water. The solution can then be pumped directly into the baker tank or pumped into the line carrying the discharged chlorine solution from the well to the storage tank. This will facilitate the blending and speed up the dechlorination process. Proper control must be used in order to assure complete removal of the chlorine and to prevent excessive use of the sodium sulfite. Indicator paper for chlorine measurement is available or field adaptable test kits can be obtained from several manufacturers for the measurement of the chlorine concentration (see Chapter 12). Two levels of testing will be required. A high level able to measure the chlorine in the well discharge and a low level usually 0 to 5 mg/l to measure the discharge and assure that no chlorine remains in the discharged water.

Since the 5,000-gallon baker tank will be used for the chlorine removal, you should measure the chlorine level once the tank is full and calculate the needed sodium sulfite according to the following formulae.

.015 lbs. of sodium sulfite for every mg/l chlorine per 1,000 gals. of water

Example: 5,000 gals. at 160 mg/l chlorine

$$.015 \text{ lbs.} \times 160 \times \frac{5{,}000}{1{,}000} = 12 \text{ pounds of sodium sulfite}$$

Use the above formulae after checking the chlorine level until the water being pumped from the well has less than 1 mg/l chlorine

Review

1) We used 12% phosphoric acid because a moderate hardness was available and in addition the well had been loosing capacity for an extended period. This may indicate a more severe blockage.

2) NuWell-310 was chosen to supply adequate dispersant chemistry for increased dissolution of minerals and to assure good wash out from the well. The higher level of NuWell-310 is also required for the iron oxidizing bacteria noted.

3) Soda ash provides excellent neutralization with buffering control and is not as difficult to handle as the sodium hydroxide or potassium hydroxide.

4) Sodium hypochlorite and NuWell-410 was selected for the disinfection. The sodium hypochlorite is much more soluble than the calcium hypochlorite and would penetrate the aquifer more easily without hardness fall out. NuWell-410 will insure proper pH control during chlorination and will provide additional penetration and calcium control.

5) Sodium sulfite is a good chlorine neutralizer and readily available. It can be purchased in 50 pound bags.

CHAPTER 14 Case Histories

This chapter is made up of experiences garnered over the years and chosen because each was a unique problem. They are related here because some of us learn more by example than we do from reading all the technical data.

No. 1: Oil Contaminated Potable Water Well

This was a gravel packed well in an unconsolidated river alluvium in which the aquifer had become contaminated with oil from a damaged pump. The damage had gone unnoticed for some time and considerable oil had been lost. The well had been operated for five years until testing positive for coliforms. Subsequent chlorination procedures had not been successful. Our laboratory had performed a complete chemistry and biological workup including a solid sample analysis.

Well Specifications:

Well Diameter: 12 inches   Total Depth: 215 feet
Static Water Level: 58 feet

The following observations were made from the laboratory tests:

Analysis indicated high levels of alkalinity and hardness. Specific ion concentrations of calcium, sulfate and silica were high and within range for mineral precipitation. A moderately high oxidation reduction potential was measured as well as a positive Saturation Index.

Biological analysis of the sample identified a moderately large sized bacterial population for an aquifer sample. The sample also identified an anaerobic bacteria population of 15% of the total bacterial count. Testing for sulfate-reducing bacteria was negative. *Pseudomonas* and *Ralstonia* species bacteria were identified; these are slime forming bacteria noted for high production rates of biofilm.

The video log showed various forms of incrustation both behind the screen and on the screen surface in the deeper section of the well. Evidence of anaerobic growth was quite noticeable at the 180 to 195 foot level. Reddish mud-like deposits below this mark were evident. Congealed and/or partially degraded oil also appeared on the screen in several locations.

Deposit Sample:

Appearance: Mud-like, grainy, moist, dark brown with outside layer being orange to reddish-brown.

Microscopic of sterile wash: Moderate to heavy bacterial activity (diverse population), some smaller fast moving protozoa, numerous unattached small *Gallionella* stalks, moderate amount of iron oxide.

Interpretations:

Chemical analysis of the water identified several parameters in the range favoring mineralogical precipitation. Specifically, the calcium and sulfate ion concentrations indicate a high potential for calcium sulfate precipitation in addition to possibly a high calcium carbonate deposition. The oxidation reduction potential and Saturation Index also indicated an environment favoring mineral formation.

The introduction of food grade oils due to pump malfunction had likely increased bacterial populations by providing an excellent nutrient source as evident in the video surveys. The continued occurrence of positive coliform testing is most likely a result of the anaerobic population which we have found often harbor coliform bacteria.

The solid sample submitted showed a considerable amount of residual oil partially degraded. In addition, there were significant amounts of iron oxide, iron sulfides and carbonates. When subjected to acid wash, the oil deposit was non reactive. Success was noted with the use of a caustic wash of the sample.

Recommendations Given:

To address bacterial effects on production quality for this well, a three-step rehabilitation program was recommended. The first step was a total volume caustic wash utilizing potassium hydroxide, NuWell-320 (a dispersant for use with caustic), and a surfactant, NuWell-400. This step was necessary to address the masses of carbonized oil and anaerobic growth within the well column and gravel pack. Special attention was recommended in the 175 to 195 foot depths as the well video showed significant accumulations within that area.

Secondly, a chemical rehabilitation with an acid and a biodispersant, NuWell-310, was recommended to address the remaining iron oxides, iron sulfides, and carbonates. A total volume pH controlled chlorination followed the rehabilitation as the third step for disinfection. The total volume method is stressed due to the severity of the bacterial influence in the casing, gravel pack and formation boundaries. This is necessary to deliver sufficient chemistry to all affected regions in effect flooding the column, borehole, and immediate formation.

Action Taken:

The caustic wash was performed using 3% potassium hydroxide, 1.5% NuWell-320, and 0.2% NuWell-400. The biodispersant was necessary to aid the caustic in breaking down the bacterial exopolymer and its removal from the well. NuWell-320 also prevented the severe mineral precipitation present under high pH conditions. The NuWell-400 is a surfactant and was used to increase penetration and help solubilize the oil residue.

The acid cleaning utilized a 10% concentration of phosphoric acid along with a 3% solution of NuWell-310. In order to target the anaerobic growth located deep within the well, gravel pack and aquifer, aggressive surging was required. Thirty minutes to one hour was recommended for every 20-foot section of casing and screen. After working the well the first day the solution was allowed to stand overnight. The well was worked again the following day for an additional six to eight hours before pump out. The area at the 175 to 195 foot depth was isolated with packer, flooded with chemistry, and aggressively surged. This area appeared to have considerable anaerobic growth requiring extra attention. The pH was maintained at 3.0 or below during the entire acid cleaning process to maximize rehabilitative efforts.

Following the chemical rehabilitation a pH-controlled chlorination was performed for final disinfection. This was scheduled after all debris from the chemical rehabilitation had been evacuated from the well and the well has been returned to "normal operating conditions." It was carried out at a pH of 6.0 to 7.0 with an approximate chlorine level of 200 mg/l. As with the caustic wash, the NuWell-400 was used to increase the penetration of the chlorinating solution. NuWell-310 was used to buffer the pH during chlorination and provide additional dispersency during the disinfection process.

The following chemicals were used and the sequence of chemical addition given for the various steps:

Caustic Cleaner (blend)

Potassium Hydroxide (45%) . . . . 70 gallons
NuWell-320 . . . . 20 gallons
NuWell-400 . . . . 3 gallons
Water . . . . 1,407 gallons
Total . . . . 1,500 gallons

Acid Cleaning Chemical (blend)

Phosphoric Acid (75%) . . . . 125 gallons
NuWell-310 . . . . 40 gallons
NuWell-400 . . . . 1.5 gallons
Water . . . . 1,333.5 gallons
Total . . . . 1,500 gallons

Chemicals for Chlorination (blend)

Sodium Hypochlorite (10%) . . . . 12.0 gallons
NuWell-310 . . . . 14.0 gallons
Water . . . . 5,074 gallons
Total . . . . 6,000 gallons

All chemistries were tremied into place in the following sequence: (The chlorine solution was prepared in 4-1,500 gallon batches)

At or just below static water level . . . . . 100 gallons
At 120 feet below ground surface . . . . . . 200 gallons
At 150 feet below ground surface . . . . . . 200 gallons
At 160 feet below ground surface . . . . . . 800 gallons
At 200 feet below ground surface . . . . . . 200 gallons

Results:

Following the chlorination and pump out, the well was put back into service. Subsequent testing was negative for coliform and certification tests since have all been negative.

No. 2: Coliform Positive New Wells

These were two 1,330 feet open borehole wells which had pressure grouted 18 inch casing to 1,150 feet. They had been drilled about eight months before, but due to power constraints in the area no pumps had been installed, although both wells had been developed. Once power had been brought to the area, pumps were installed and both wells were chlorinated in anticipation of bringing them on line and turning them over to the owner. Two more subsequent chlorinations, one including a pH control application, were unsuccessful and both wells remained positive for coliform bacteria. Samples were then sent to our laboratory for evaluation.

Well Specifications:

Well Diameter: 18 inches Well Depth: 1,330 feet
Static Water Level: 460 feet

The following observations were made from the laboratory tests:

Microscopic:

Well No. 3: Clean sample, no bacterial activity, no sheathed or stalked bacteria.

Well No. 4: Clean sample, no bacterial activity, no sheathed or stalked bacteria.

Chemical analysis presented relatively clean samples from each well. Iron and manganese concentrations were low, if detected. Calcium and magnesium appeared to be within normal background levels. Mineral scale development should have been expected over time but was dependent on operating conditions.

The biological analysis by plate count and ATP analysis indicated a significant bacterial presence especially for a new well. Anaerobic growth studies showed large percentages present in addition to positive sulfate-reducing bacterial presence. Each sample tested positive for coliform bacteria and negative for *E. coli* bacteria. *Pseudomonas fluorescence, Comamonas acidovorans,* and *Klebsiella pneumoniae* (the coliform) were all identified in each of the wells. Coliform counts from a local laboratory had been recorded at 37/100 ml for Well No. 3 and 106/100 ml for Well No. 4.

Since both wells sat idle for several months following development and before the permanent pump installation, high bacterial counts were expected. Well No. 3 sat approximately twice as long and the longer time period was reflected in its higher biological populations. During static periods both aerobic and anaerobic bacterial populations blossom and population counts rise dramatically.

The presence of sulfate-reducing bacteria in both samples (drawn from new wells) was surprising especially since wells in this area are reported to experience a high level of hydrogen sulfide gas from "old mining activity." The combination of sulfate-reducing bacteria and a high population of anaerobic organisms indicated a more involved and troublesome biological activity within the well system. This condition would cause an increased chlorine demand as well as a need for enhanced mechanical efforts to address the affected biomass. During construction a fifty-foot sump had been constructed in each well.

Recommendations Given:

Standard treatment responses for wells of this nature would include the use of a pH-buffered chlorination treatment or volume displacement chlorination. Since both of these had already failed (15,000 gallons of hypochlorite solution had been placed in each well twice, once with pH control) and we had learned of the existence of the sumps, the following program was devised. Sumps are an excellent hiding place for the anaerobic sulfate-reducing bacteria and the coliform organisms. Coliforms are facultative, that is they can live in both aerobic and anaerobic conditions; and since the more aerobic zones had been thoroughly chlorinated they were hiding in the anaerobic biofilm, most probably in the sump area (see Chapter 1).

A two-inch plastic pipe was fitted along side the pump and reached to the sump. Chlorine tablets were blown with nitrogen gas down to the sump areas and the gas was used to disrupt the heavy biofilm in this area. The treated wells were allowed to set overnight and each well was pumped hard the next day.

Results:

After waiting the required regulatory time before testing, each well was tested and reported coliform free. Each unit has been online and coliform free for approximately one year.

No. 3: Very Limited Specific Capacity after Three Development Periods Using Phosphates

A large Texas well which included a 48-inch underreamed gravel pack area had been developed with phosphates following construction and had only succeeded in getting approximately 27% efficiency (contractor's figures). Since the contract called for a minimum of 75% efficiency additional phosphate development treatments had been undertaken. Although laboratory analysis was not performed, it was decided after looking at the geology that clay content was being adversely affected by the phosphates. This occurred because of the chemistry and because of the excessive mechanical activity which was upsetting the distribution of the natural formation (three development periods).

Well Specifications:

Well Diameter: 16 inches
Well Depth: 1,400 feet
Static Water Level: 180 feet

Observations:

The clays could have been affected and disruption of the aquifer may have occurred, but also the probability existed that the mud used during construction was not being removed, especially in the large underreamed area.

Recommendations Given:

We recommended the use of NuWell-220, a polymeric dispersant extremely active as a mud control agent. The product was used at one gallon for every 500 gallons of standing water, times 1.5 and developed as before.

Results:

The results were a 50% increase in the expected efficiency (well contractor's figures). The contractor wanted to try again considering the success but a second try yielded only a one percent increase. This was expected due to the very high activity this product exhibits against mud and clays. Use applications in the original testing had shown very little improvement with subsequent usage. (Efficiency figures provided by the contractor were based on the average capacities observed in the area and the expected yield per contract.)

No. 4: Potable Well Water Distribution System with "Floating Black Spots"

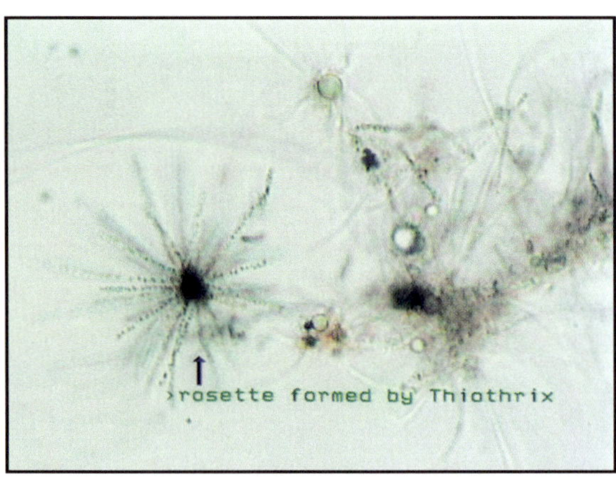

Figure 14.1. Thiothrix

A city in the Southwest sent in a water sample with what appeared as floating or suspended black spots. Microscopic analysis showed a typical rosetta formation of the bacteria *Thiothrix* (Figure 14.1). A well water sample was also submitted for analysis.

Well Specifications:

Well Diameter: 16 inches Well Depth: 1,200 feet
Static Water Level: 320 feet

The following observations were made from the laboratory tests:

The water analysis showed very biocongested water with excessive levels of bacteria as noted by both the plate count and the ATP analysis. Considerable anaerobic organisms (sulfate-reducing bacteria) were noted in addition to the *Thiothrix* bacteria. The *Thiothrix* are responsible for the black filamentous material in the distributed water. This organism requires reduced sulfur such as the sulfide present. The presence of sulfides in the water is due to the sulfate-reducing bacteria present. When the well sets idle, growth accelerates and the organisms accumulate in high numbers giving the dark appearance and the presence of numerous filamentous clumps.

Since *Thiothrix* attach by hold fasts they are able to set up dense biofilm, not only on surfaces in the well but in piping throughout the distribution system, and slough off the characteristic clumps as the water flows.

Recommendations Given:

There are only two products NSF certified specifically for cleaning wells and distribution piping that were found effective in heavy biofilm removal. Those products are NuWell-310 manufactured by Johnson Screens and QC-21 manufactured by Layne Christensen Co.

A low acid concentration of 3 to 4% and a 3% concentration of either NuWell-310 or QC-21 products should be used. If pumping the well for an extended period confirms the unusual heavy bio load, then two applications of chemistry should be applied. The first application should be allowed in the well overnight followed by surging with a surge block and then discharged. Next a new mixture of acid plus biodispersant should be added and surged into place. This should be left in overnight with surging prior to discharge the next day. Cleaning of the distribution system can be carried out with a low level acid concentrate and a 0.5 to 1.5% concentration of the biodispersant, however, it has been our experience that once the sulfide source has been eliminated the *Thiothrix* will disappear.

Results:

The first year the distribution piping only was cleaned because the water superintendent didn't believe our report. Once the problem returned the new superintendent commissioned a well contractor to follow our recommendations and the well and system have remained clean for a number of years.

No. 5: Potable Water Well with Pathogenic Bacteria Contaminant

A screened, gravel-packed, and underreamed well was producing colored water. Since it was a backup well for a potable water supply the city asked that the colored water be investigated. The contractor sent two samples into our laboratory. One sample was taken just after startup, but after the down pipe had emptied. The well had set idle overnight per our instructions. The second sample was collected after the well had been pumped for three hours. This represented an aquifer sample and the first sample was considered the casing sample. This is a repeat of a similar well contamination seen several years ago in a California community well.

Well Specifications:

Well Diameter: 8 inch casing and 4 inch screen
Well Depth: 960 feet Static Water Level: 149 feet

The following observations were made from the laboratory tests:

Chemical analyses of the casing and aquifer samples showed excessive dissolved solids and alkalinity, however no hardness was detected. Specific ion concentrations were low including iron, manganese, and sulfate. The calculated Saturation Indexes were negative and the oxidation reduction potential readings indicate a moderately oxidative condition.

Biologically the samples exhibited an excessive bacterial ATP count and heavy heterotrophic plate growth. *Flagellates* detected in the casing sample did add somewhat to the ATP count, however the heterotrophic plate count would not have been affected. Testing for sulfate-reducing bacteria and coliform growth was negative. Identifications were made of *Chromobacterium violaceum* and *Pseudomonas viridilivida* as dominant populations in both samples.

The bacterial ATP and heterotrophic plate counts were indicative of a very large bacterial population representative of high growth and activity. The aquifer count was well above average for a pumping sample, a reflection of heavy growth in the gravel pack and formation area around the well. The identified *Pseudomonas* specie (casing and aquifer) was a heavy biofilm producer which contributed significant biomass accumulation throughout the well system.

The bacterium *Chromobacterium violaceum* is a highly pigmented bacterium which is most likely the source of the color problem. (This bacteria is also an occasional pathogen for humans resulting in pyogenic or septicemic infections.)

Recommendations Given:

Due to the presence of the *Chromobacterium* specie and the heavy overall bacterial growth, this well should first be thoroughly cleaned with acid and biodispersant chemistry to remove the heavy biofilm as this material will not be sufficiently removed through typical chlorination. After chemical cleaning of the well, follow with a pH-controlled chlorination. The chemical rehabilitation and chlorination procedures should each be performed over separate two-day periods allowing for the well to stand overnight with the chemistry in place. The drop pipe and all equipment utilized in the operation should be thoroughly disinfected before re-installation in the well to prevent recontamination.

As soon as possible the well should be disconnected from the distribution system to limit the potential for contamination of connected piping. Following the cleaning, submit a new sample after a two-week period for analysis to recheck for the presence of the *Chromobacterium* organism. Consideration should be given to testing the distribution system for the presence of the organism if this well has been utilized for production.

Based on the laboratory analysis, chemical rehabilitation of the well is recommended utilizing the QC-21 or NuWell-310 biodispersants at a full 3% concentration to provide for dissolution and suspension of the organic accumulation and for its removal to waste. Phosphoric acid is recommended at a 6% concentration to maintain the pH of the downhole solution at a level sufficient to maximize the efficiency of the cleaning process.

To improve rehabilitative efforts the casing and screen should be brushed with evacuation of debris prior to the addition of the chemistry. To address the heavy bacterial growth located in the gravel pack and formation areas aggressive surging will be necessary. Depending on depth, 30 minutes to one hour would be recommended for every 20 foot section of screen. The casing area should also be surged, although not as long, to provide cleaning of biofilm from this area. After working the well the first day the solution should be allowed to stand overnight. The well should be worked again the following day for an additional six to eight hours before pump out. Check the pH after the first day and if above 3.0 add additional acid before letting set overnight. With chemical rehabilitation of this type it is very important to maintain the pH at 3.0 or below during the entire cleaning process to maximize rehabilitative efforts. Be sure to neutralize pump out above ground and dispose of according to regulatory requirements.

Based on the information provided on the well data sheet, a 2,000-gallon treatment solution will be required and includes the following volumes:

110 gallons of food grade phosphoric acid (75% A.I.)
55 gallons of biodispersant (QC-21 or NuWell-310)

Chlorination:

To provide for the most effective chlorination procedure, a chlorine solution should be produced which is equal to four times the standing well volume (6,000 gallons) to effectively flood the entire well and penetrate the surrounding gravel pack and immediate formation. This will require the use of the biodispersant for pH control and the use of 200 mg/l chlorine.

Biodispersant (QC-21 or NuWell-310). . . . 18.0 gallons
Sodium Hypochlorite (10%). . . . . 12.0 gallons

After addition of the chlorine solution to the well, the same mechanical activity should be applied as utilized for the acid/ biodispersant cleaning. The high amount of dispersant required is due to the excessive water alkalinity.

Results:

The contaminated well is presently undergoing treatment, however a similar California well was cleaned successfully several years ago and has not had any reappearance of the problem organism.

No. 6: Pressure Relief Wells in a Dam Installation

Figure 14.2. Collar for use in cleaning interior pressure relief wells

In the Southeast a number of wells located in several major dams were in need of cleaning. They were completely plugged with a heavily incrusted biofilm and mineral complex. The wells were located in galleries or passages in the dam interior. The only way they had been previously cleaned was to take a drill rig apart and reassemble it in the chamber where the wells were located.

Recommendations Given:

We recommended that a 55-gallon open drum of cleaning solution be prepared with 5% NuWell-110 and 3% NuWell-310. The NuWell-110 was blended into 40 gallons of water and after thorough mixing the NuWell-310 product was added. This mixture was pumped into the well through a one inch plastic pipe inserted down to the blockage. The solution and dissolved minerals traveled back up the casing and were returned to the drum through a collar which sealed the top of the casing. This collar held the down pipe in place and diverted the returned fluid to the drum. The recirculated solution was drawn from higher up in the drum to allow solids to fall to the bottom. The pH of the solution was checked often to insure that a pH of 3.0 was maintained in the circulating cleaner.

Results:

Considerable amount of solids were removed with the circulating flow and this presented a handling problem in the start up of each well. The solids also continued to neutralize the cleaning solution before they were removed. This problem resulted in a number of fluid changes in the first part of cleaning for each well. Once this was overcome the process for each well ran smoothly. Cleaning was excellent and was completed in record time and at a minimum of cost.

No. 7: The Mysterious Chlorine Loss

Several wells in a river alluvium in the South were being rehabilitated and the consultant had used our services in determining the type of chemical and best procedures for the chemical cleaning. All had gone well and following the pump out of the acid cleaning chemistry the wells were allowed to set over the weekend before chlorination. Chlorination had been left up to the contractor.

Well Specifications:

Well Diameter: 8 inches   Well Depth: 50 feet
Static Water Level: 6 feet

The following observations were made:

The wells had a standing well volume of 115 gallons which when multiplied by four gave a required disinfectant volume of 460 gallons. If they had wanted to treat the well at the 200 mg/l level of chlorine, it would have taken approximately one gallon of a 10% sodium hypochlorite solution. They had planned to use five gallons of sodium hypochlorite per well.

Following the first application of five gallons of the hypochlorite and the second application of 25 gallons of sodium hypochlorite they called to let us know that they had a very high chlorine demand as chlorine tests showed very low levels of hypochlorite. Their plans were to dump ten five-gallon buckets into each of the wells. They assured us that the pH of the water was normal and that they had thoroughly pumped the wells to remove the acid cleaner.

Recommendations Given:

Check the pH of each of the wells before they add sodium hypochlorite to the well.

Results:

The first well was checked and the water was at a pH of 3.0. Apparently acid cleaner had gotten caught back in the formation and was only gradually being released. Sodium hypochlorite below a pH of 5.0 will release chlorine gas which escapes from the water and is either lost to the atmosphere or to the upper formations of the well. Subsequent pump out raised the pH to the level before acid cleaning and chlorination proceeded without a problem.

No. 8: An Example of Well Monitoring and Preventative Maintenance

An open borehole well had been used as a water source for an industrial plant for which a very high grade of water purity was required. The well water was filtered twice before entering the plant and a final time at each station. The point-of-use filters were being changed more often and eventually were not lasting more than two hours. A slimy material was noted in the distribution lines through out the plant.

Well Specifications:

Well Diameter: 8 inches   Well Depth: 407 feet
Static Water Level: 15 feet

The following observations were made from laboratory tests:

Samples taken from the well and several locations in the plant showed an excessive level of *Gallionella ferruginea* together with substantial quantities of iron oxyhydroxide. A sample collected from the well after extended pumping also checked positive for the iron oxidizer bacteria. All well analysis showed high levels of alkalinity and hardness along with a positive Saturation Index. While the first water sample showed a total iron level of 6.1 mg/l, extended pumping showed only 2.1 mg/l. No soluble iron (ferrous iron) was detected.

Recommendations Given:

As requested, recommendation was given to chemically clean the distribution system and the well. We recommended using 10% concentration of food grade phosphoric acid and a 3% concentration of NuWell-310, as manufactured by Johnson Screen, for the well cleaning. The following instructions were sent.

Chemical Requirements:

For safety purposes, especially if recirculation and jetting techniques are utilized, phosphoric acid 75% active food grade should be used for iron and carbonate removal. NuWell-310 should be used for mineral dispersancy and effective solubilizing of the biofilm. The phosphoric acid will be required at 10% of the total well volume and NuWell-310 at 3%.

Calculations:

Well volume = 407 ft. − 15 ft. (SWL) = 392 feet

SWV = 392 ft. x 2.613 gal./ft. = 1,025 gallons

Total Treatment Volume (TTV) = SWV x 1.5

1,025 gals. x 1.5 = 1,537 gallons

Phosphoric acid 75% food grade:

1,537 x 8.3 = 12,757 lbs. x 10% = 1,275.7 pounds

$$1{,}275.7 \div 13.1 \text{ lb./gal.} = 97.4 \text{ gals.} \times \frac{100}{75} = 130 \text{ gallons}$$

NuWell-310 acid enhancer:

12,757 lbs. x 3% = 382.7 ÷ 10 lb./gal. = 38.2 gallons

SWL = Static Water Level *SWV = Standing Well Volume*

The volume of chemical mix necessary to prepare for adequate treatment of the well should equal approximately 40% of the total well volume or 600 gallons. Using an 800 to 1,000 gallon capacity tank, blend the 130 gallons of phosphoric acid to 430 gallons of water. Be sure to ***blend the acid to the water*** and *not* water to acid. To this mixture add the 40 gallons of NuWell-310.

Application Method:

The following procedure should be followed to prepare the well for chemical addition and subsequent cleaning:

1) Remove the pump from the well.

2) Wire brush any casing present unless video review shows a clean surface.

3) After brushing evacuate any debris from the well starting at the static water level and working downward. Spend 10 to 15 minutes at each 20 feet interval to loosen as much debris as possible. Be sure to evacuate the bottom of all debris.

4) Add the chemical blend to the well by placing approximately 75 gallons at each 50 feet interval beginning 50 feet off the bottom and finishing at the static water level.

The cleaning process will consist of recirculation of the cleaning solution through a jetting tool to direct jets of high-pressure spray at the borehole walls. The jetting action should incorporate a minimum of 30 to 45 minutes of jetting action at each 50 feet interval. The pH of the solution should be checked periodically in order to maintain a pH of 3.0 or less in the recirculating cleaner. If the pH rises above 3.0 add 30 gallons of additional 75% phosphoric acid food grade to the solution. Repeat this each time the pH rises.

After the jetting has been completed allow the well to stand over night. Make sure the pH is at 3.0 or below. Repeat the jetting the following day spending 20 to 30 minutes at each 50 feet interval. Once the jetting procedure is complete, evacuate the well starting just below the static water level and moving downward. Spend at least ten minutes at each 50 feet interval. Check the pH. If the pH returned to normal and total evacuation equaled 50,000 gallons, the well can be prepared for chlorination.

The pump and all column piping must be thoroughly scrubbed with a mixture of 10% phosphoric acid and 3% NuWell-310 or steam cleaned prior to reinstallation in the well. Cleaning can be performed using the cleaner solution used for recirculation in the well if cleaner is not lost so that a full 600 gallons remains for well cleaning. If additional cleaner is required, a blend for each 100 gallons required can be mixed accordingly:

75% Food Grade Phosphoric Acid .	8.5 gallons
NuWell-310	3.0 gallons
Water	88.5 gallons
	100.0 gallons

Be sure to ***add the acid to the water*** and not the water to acid. Always add the NuWell-310 to diluted acid.

While phosphoric acid is much safer to use, acceptable chemical goggles, outerwear, and respiratory masks should be provided for personnel involved in the cleaning.

Chlorination should be performed following evacuation of the acid cleaner and before the well is put back into service. The jetting tool would also be useful in applying the agitation necessary for a good chlorination.

Chlorine Requirements:

Use a 200 mg/l (0.02%) concentration of sodium hypochlorite. The cleaning chemistry can be prepared utilizing the same mixing tank. Nu-Well 310 should be used to control the pH during chlorination.

Calculation:

Total Disinfectant Volume (TDV) = TTV x 2

TDV = 1,537 gals. x 2 = 3,074 gallons

$$\text{Hypochlorite gallons} = \frac{\text{TDV} \times \text{\% Chlorine Desired}}{\text{\% active hypochlorite solution}}$$

200 mg/l chlorine = .02%

$$\frac{3{,}074 \times (.02\%)}{10\%} = 6.1 \text{ gallons of 10\% hypochlorite solution}$$

NuWell-310 will be required at approximately one gallon for every 2,000 gallons of TDV or a total of 1.5 gallons.

Procedure:

1) If only the 1,000-gallon capacity tank is available three, batches of chlorine chemistry needs to be prepared. Add approximately 950 gallons of water to the mixing tank.

2) Add one-half gallon of NuWell-310 to the tank and mix briefly.

3) Check the pH and add additional NuWell-310 until the pH is 4.5 to 5.0.

4) Add the 2 gallons of the 10% sodium hypochlorite to the conditioned water and blend or mix immediately.

CAUTION: Addition of the hypochlorite will momentarily release small amounts of chlorine gas! This must be performed in the open and a respirator worn by the person adding the chlorine solution. The sodium hypochlorite is a very alkaline product and will raise the pH immediately.

Once the pH is above 5.0 no chlorine is released. The target pH or most efficient pH for well chlorination is a pH of 6.5. At this level more than 95% of the chlorine is in the hypochlorous acid form over 100 times more effective as a disinfectant.

5) Add the chlorine solution to the well using the same procedures as used for the acid cleaning solution. Repeat the above procedure until a total of 3,000 gallons of solution has been added to the well.

6) Agitate the solution with the use of the jetting system recirculating to completely wash the well. Cover the entire length of the well with the jetting action. Let stand for four hours or as required to meet your regulatory contact periods.

7) Evacuate the well until chlorine levels reach less than 50 mg/l. At this point the pump can be reinstalled. Pump the system to waste until chlorine levels reach one mg/l or less for at least one hour of pumping. The system can then be returned to operation.

Neutralization of Discharge:

If local regulatory conditions allow environment release of well cleaning discharge, then provisions for acid neutralization and chlorine reduction should be made.

Results:

The recommendations were followed in cleaning the well, however the distribution system was not cleaned. The filter systems had been completely rehabilitated or replaced. Before the pump was installed a plastic pipe line was permanently installed near the well bottom. This was recommended to facilitate the addition of light concentrations of acid and biodispersant to control any reappearance of the *Gallionella.*

Monitoring samples have been subjected to analysis for *Gallionella,* sulfate-reducing bacteria, total bacteria (ATP count), TDS, and pH monthly for the last couple of years. Twice during that period significant counts of *Gallionella* were observed and total bacterial counts rose sufficiently to require the addition of the light wash out. Each occasion resulted in a good removal and what appeared to be control of the iron oxidizer population.

Weather problems and holiday celebrations pushed the distribution system cleaning off for 30 days. Several checks of the system before the scheduled cleaning found no *Gallionella* present and on-site filters were being replaced only on scheduled maintenance. The system was not cleaned and there has been no reoccurrence of the problem. The primary filter systems have also operated with no apparent or excess blockage. Laboratory testing, particularly bacterial monitoring, can be helpful in setting up a preventative maintenance program.

APPENDIX A Potential Water Supply Pathogens

Table A.1. Human viruses found in waster materials that enter water

Viruses	Nucleic Acid	Disease(s)	Waste(s)
Adenoviruses			
Human adenovirus	DNA	Actue respiratory, pharyngitis, acute hemorrhagic cystitis	Wastewater
Enteric adenovirus	DNA	Gastronenteritis	Wastewater
Caliciviruses			
Calcivirus	RNA	Gastronenteritis	Wastewater
Norwalk virus	RNA	Gastronenteritis	Wastewater
Coronaviruses			
Enteric coronavirus	RNA	Intestinal disorders	
Orthomyxoviruses			
Influenza virus	RNA	Influenza	Human, swine and fowl wastes
Picornaviruses			
Coxsackievirus A	RNA	Miningitis, herpangia, common cold	Wastewater
Coxsackievirus B	RNA	Mycarditis, pleurodynia, rash, meningitis, paralysis	Wastewater
ECHO virus	RNA	Paralysis, diarrhea, meningitis	Wastewater
Hepatitis A virus	RNA	Infectious hepatitis	Wastewater
Poliovirus	RNA	Poliomyelitis	Wastewater
Reoviruses			
Reovirus	RNA	Respiratory, gastroenteritis	Wastewater
Rotavirus	RNA	Infantile diarrhea	Wastewater

Table A.2. Pathogenic bacteria found in waster materials that enter water

Species	Disease(s)	Waste(s)
Acinetobacter calcoaceticus	Nosocomial	Water, human skin and mouth
Aeromonas hydrophila	Septicemia, wound infections, diarrhea	Water (fresh and estuarine)
Aeromonas sobria	Septicemia, wound infections, diarrhea	Water (fresh and estuarine)
Aeromonas caviae	Septicemia, wound infections, diarrhea	Water (fresh and estuarine)
Bacteroides fragilis	Intraabdominal abscesses	Animal feces
Bacteroides melaninogenicus	Orofacial	Human mouth and feces
Brucella spp.	Brucellosis	Animal feces, urine, and milk
Campylabacter fetus	Septicemia	Animal feces
Campylobacter jejuni	Enteritis	Animal feces
Chromobacterium violaceum	Septicemia and diarrhea	Soil and water
Citrobacter spp.	Nosocomial	Water
Clostridium botulinum	Botulism	Soil, sediment, and fish
Clostridium difficile	Pseudomembranous colitis	Vagina and gastrointestinal tract
Clostridium perfringens	Gangrene, wound abscesses, and food poisoning	Animal feces
Clostridium sporogenes	Gangrene	Soil and animal feces
Clostridium tetani	Tetanus	Soil and animal feces
Coxiella burnetii	Q fever	Milk and animal wastes
Enterobacter spp.	Nosocomial	Wastewater
Erysipelothrix rhusiopathiae	Erysipeloid	Animal feces and fish slime
Escherichia coli	Gastroenteritis	Wastewater
Flavobacterium meningosepticum	Nosocomial, meningitis	Freshwater
Francisella tularensis	Tularemia	Rodents and freshwater
Fusobacterium necrophorum	Liver and soft tissue abscesses	Wastewater and animal feces
Klebsiella pneumoniae	Pneumonia, bacteremia, and nosocomial	Water, feces, soil and plants
Legionella pneumophila	Legionnaires' disease	Freshwater, cooling tower water, and hot water tanks
Leptospira interrogans	Leptospirosis	Urine
Listeria monocytogenes	Listeriosis	Soil and feces
Morganella morganii	Urinary tract and nosocomial	Water, feces, and decaying animals
Mycobacterium tuberculosis	Tuberculosis	Wastewater
Mycobacterium marinum	Swimming pool granuloma	Water and fish
Plesiomonas shigelloides	Gastroenteritis	Water, fish, and aquatic animals
Proteus spp.	Urinary tract and nosocomial	Water, feces, and decaying animals
Pseudomonas aeruginosa	Burn, wound, corneal, ear, urinary, lung, skin, and gastrointestinal tract	Water, wastewater, plants, sediment, and fish
Pseudomonas pseudomallei	Meliodosis	Water and soil
Salmonella typhi	Typhoid fever	Wastewater
Salmonella enteritidis	Gastroenteritis and septicemia	Wastewater, animal wastes and feed, and compost
Serratia marcescens	Nosocomial	Water, plants, insects, and feces
Shigella boydii	Bacillary dysentery	Primate feces and wastewater
Shigella dysenteriae	Bacillary dysentery	Primate feces and wastewater
Shigella Flexneri	Bacillary dysentery	Primate feces and wastewater
Shigella sonnei	Bacillary dysentery	Primate feces and wastewater

Table A.2. Pathogenic bacteria found in waster materials that enter water *(continued)*

Species	Disease(s)	Waste(s)
Staphylococcus aureus	Abscesses and food poisoning	Mammalian skin and ocean water
Streptococcus faecalis	Endocarditis	Animal feces
Vibrio Alginolyticus	Wound infections	Ocean water and aquatic animals
Vibrio cholerae	Asiatic cholera	Wastewater, shellfish, and saltwater
Vibrio parahaemolyticus	Gastroenteritis	Saltwater and shellfish
Vibrio vulnificus	Septicemia and wound infection	Oysters and seawater
Yersina enterocolitica	Gastrointestinal, acute mesenteric	Water, milk, mammalian alimentary canal

Table A.3. Pathogenic fungi associated with waste material that enter water

Fungus	Disease(s)	Waste(s)
Aspergilus fumigatus	Aspergillosis	Decaying vegetation, especially grains
Candida albicans	Candidiasis	Animal feces
Cryptococcus neoformans	Crytococcosis	Pigeon and bird feces, cellar dirt
Geotrichum candidum	Geotrichosis	Tomatoes, fruits, dairy products
Histoplasma capsulatum	Histoplasmosis	Chicken feces, bat guano

Table A.4. Waste associated cyanobacteria and eucaryotic algae pathogenics for humans

Species	Disease(s)	Waste(s)
Cyanobacteria		
Anabaena Flos-aquae	Neuromuscular poison	Eutrophication and blooms
Aphanizomenon Flos-aquae	Neuromuscular poison	Eutrophication and blooms
Lyngbya majuscula	Swimmers' itch	Seawater
Microcystis aeruginosa	Hepatomegaly and liver necrosis	Eutrophication and blooms
Oscillatoria migroviridis	Swimmers' itch	Seawater
Schizothrix calcicola	Swimmers' itch	Seawater
Eucaryotic algae		
Gonylaulax spp.	Paralytic shellfish poisoning	Unknown
Gymnodinium breve	Paralytic shellfish poisoning	Unknown
Pyrodinium monilatum	Paralytic shellfish poisoning	Unknown
Gambierdiscus spp.	Ciguatera seafood poisoning	Unknown

Table A.5. Waste associated protozoa pathogenic for human

Species	Disease(s)	Waste(s)
Mastigophora (flagellates)		
Chilomastix mesnili	Diarrhea	Primate feces
Giardia lamblia	Giardiasis	Human feces
Sarcodina (amebas)		
Entamoeba Bistolytica	Amebic dysentery	Human and other animal feces, wastewater
Dientamoeba fragilis	Mild diarrhea	Human feces
Naegleria fowleri	Primary amebic meningoencephalitis	Human feces, wastewater
Acanthamoeba spp.	Amebic meningoencephalitis	Human feces, wastewater

Table A.5. Waste associated protozoa pathogenic for human *(continued)*

Species	Disease(s)	Waste(s)
Sporozoa		
Cryptosporodium spp.	Cryptosporidiosis	Animal feces
Sarcocystis spp.	Sarcocystosis	Animal feces
Toxoplasma gondii	Toxoplasmosis	Animal feces, especially cats
Ciliata		
Balantidium coli	Balantidiasis	Animal feces, especially swine

Table A.6. Waste associated helminths pathogenic for humans

Species	Disease(s)	Waste(s)
Digenetic trematodes (flukes)		
Schistosoma baematobium	Schistsomiasis	Human feces
Schistosoma japanicum	Schistsomiasis	Human feces
Schistosoma mansoni	Schistsomiasis	Human feces
Echinostoma spp.	Diarrhea	Animal feces
Faxciola hepatica	Liver necrosis and cirrhosis	Animal feces
Paragonimus westermani	Paragonimiasis	Animal feces and crustaceans
Clonorchis sinensis	Bile duct erosion	Human feces and raw fish
Heterophyes heterophyes	Diarrhea and myocarditis	Human feces and raw fish
Cestodes (tapeworms)		
Diphyllobothrium latum	Diarrhea and anemia	Human feces and raw fish
Taeniarhynchus saginatus	Dizziness, nausea, pain, and inappetence	Human feces and raw beef
Taenia solium	Dizziness, nausea, pain, inappetence, cysticercosis	Human feces and raw pork
Echonococcus granulosis	Hydatidosis	Dog and other animal feces
Hymenolepis nana	Dizziness, nausea, pain, and inappetence	Human feces
Nematodes (roundworms)		
Trichuris trichiura	Asymptomatic to chronic hemorrhage	Human feces
Trichinella spiralis	Trichinosis	Raw or undercooked meat
Strongyloides stercoralis	Strongyloidiasis	Human feces
Necator americanus	Iron deficiency anemia and protein deficiency	Human feces
Ancylostoma duodenale	Iron deficiency anemia and protein deficiency	Human feces
Ascaris lumbricoides	Ascariasis	Human, pig, and other animal feces

APPENDIX B Commercially Available Chemicals

Acids

Hydrochloric Acid (NSF Available) (Muriatic Acid)
20.0% AI – 9.13 lbs./gal.
29.5% AI – 9.50 lbs./gal.
39.1% AI – 10.00 lbs./gal.
available in totes, 140 and 500 pound drums

Hydroxyacetic Acid (70% AI) (Glycolic Acid)
10.4 lbs./gal., available in 550 pound drums

NuWell Pellitized Acid (Formerly NuWell-100)
70.0 lbs./cubic ft., available in 4.5 and 9 pound jars,
45 and 70 lb. drums; NSF Certified;
from Johnson Screens, Inc.

NuWell Granular Acid (Formerly NuWell-110)
80 lbs./cubic ft., available in 50 pound pails; NSF certified;
from Johnson Screens, Inc.

Phosphoric Acid (NSF or Foodgrade Available)

75% AI – 13.00 lbs./gal., available in 200 and 650 pound drums

85% AI – 13.98 lbs./gal., available in 750 pound drums

Sulfamic Acid Granular

70 lbs./cubic ft., available in 50 pound bags

UNICID Granular Acid/Bullets

80 lbs./cubic ft. available in 50 pound pails; from Design Water Technologies

Caustics

Lime (Hydrated)

Available in 50 pound bags

Magnesium Hydroxide 60% active

12.3 lbs./gal., available in 670 pound drums

Potassium Hydroxide 45% active (Liquid Caustic Potash)

12 lbs./gal., available in 650 pound drums

Soda Ash (Sodium Carbonate)

Available in 50 pound bags

Sodium Hydroxide (Liquid Caustic 50%)

12.8 lbs./gal., available in 650 pound drum and totes

Biodispersants

Unicid Catalyst

9.3 lbs./gal., available in 1, 5, 30, and 55 gallon containers; from Design Water Technologies

BioAcid Dispersant (Formerly NuWell-310)
10 lbs./gal., available in 1, 5, 30, and 55 gallon containers; NSF Certified; from Johnson Screens, Inc.

BioCaustic Dispersant (Formerly NuWell-320)
9.5 lbs./gal., available in 1, 5, 30, and 55 gallon containers; from Johnson Screens, Inc.

QC-21 Well Cleaning Product
9.5 lbs./gal., available in 5, 30 and 55 gallon containers; NSF Certified; from Layne Christensen Co.

Clays, Bentonites & Silt Active Polymers

MudBuster
11.0 lbs./gal., available in 1, 5, 30, and 55 gallon containers; from Design Water Technologies

Clay Dispersant (Formerly NuWell-220)
10.5 lbs./gal., available in 1, 5, 30, and 55 gallon containers; NSF Certified; from Johnson Screens, Inc.

pH Control Chemistry

ChloraPal
9.2 lbs./gal., available in 1, 5, 30, and 55 gallon containers; NSF Certified; from Design Water Technologies

Chlorine Enhancer (Formerly NuWell-410)
9.3 lbs./gal., available in 1, 5, 30, and 55 gallon containers; NSF Certified; from Johnson Screens, Inc.

Surfactants

Surfactant (Formerly NuWell-400)
9.4 lbs./gal., available in 1, 5, 30, & 55 gallon containers; NSF Certified; from Johnson Screens, Inc.

Triton 100
8.9 lbs./gal., available in 5 & 55 gallon containers from Rohm & Haas

Chlorine Neutralizers

Ascorbic Acid, Powder/Crystals
Available in 25 kilo box

ChorOut (Formerly NuWell-500)
75 lbs./cubic ft., available in 10 pound containers; from Johnson Screens, Inc.

Sodium Sulfite, Powder
Available in 50 pound bags

Sodium Thiosulfate (anhydrous crystals)
Available in 50 pound bags

Chlorine Products

Sodium Hypochlorite 5.25% active
9.0 lbs./gal., available as household bleach

Sodium Hypochlorite
10.0% active 9.6 lbs./gal.
12.0% active 10.0 lbs./gal.
available in 5, 15, and 54 gallon containers

Calcium Hypochlorite 65% active, Powder/Crystals

Available in 1, 5, 25, 100 pound containers

Sterilene, Granular

Available in case lots, 50 pound containers; from Design Water Technologies

APPENDIX C Calculations

Table C.1. Useful Equivalents

1 milliliter	1 cubic centimeter (cc)
1 fluid ounce	29.57 milliliters
1 pint	473.2 milliliters
1 quart	946.3 milliliters
1 gallon	3785.2 milliliters
1 ounce	28.34 grams
1 pound	453.59 grams
1 kilogram	2.20 pounds
1 cubic foot	7.48 gallons (US)
1 cubic meter	1,000 liters
1 cubic meter	264.20 gallons (US)
1 cubic yard	202.0 gallons (US)
1 inch	2.54 centimeters
1 foot	0.30 meters
1 yard	0.91 meters
1 meter	3.28 feet
1 gallon	8.34 pounds
1 gpm	0.06 liters per second
cubic foot / minute	0.47 liters per second

Table C.2. Pipe capacity

Diameter Pipe/Hole (inches)	Gallons per Foot
1.000	0.041
1.500	0.092
2.000	0.163
2.500	0.255
3.000	0.367
3.500	0.500
4.000	0.653
4.500	0.827
5.000	1.021
5.500	1.235
6.000	1.470
6.500	1.725
7.000	2.000
7.500	2.296
8.000	2.613
9.000	3.307
10.000	4.082
12.000	5.879
14.000	8.002
16.000	10.451
18.000	13.227
20.000	16.330
22.000	19.759
24.000	23.515
28.000	32.006
30.000	36.742
36.000	52.908
48.000	94.059
60.000	146.968
72.000	211.633
84.000	288.056
96.000	376.237

Table C.3. Total Treatment Volume (TTV) Calculation

Standing Well Volume (SWV)

SWV = gal/ft (well diameter) x feet of water

Annulace Volume (AV)

AV = borehole gals/ft – casing gals/ft x feet of water x porosity

Total Treatment Volume (TTV)

TTV = SWV + AV

Example: Borehole Diameter = 24 inches
Casing/Screen Diameter = 16 inches
Water column = 100 feet
Porosity of Gravel Pack = 25%

SWV = 10.45 x 100 ft = 1,045 gallon

AV = 23.51 - 10.45 x 100 x 0.25 = 326.5 gallon

TTV = 1,045 gals + 326.5 gals = 1371.5 gallon
1,371.5 x 8.34 = 11,438 pounds of water

Table C.4. Total Treatment Volume TTV (Approximated)

Gravel pack (not over sized or underreamed) and open borehole wells

SWV = gal/ft x ft of water

For wells with 6 inches or less casing diameter

TTV = 2.0 x SWV

For wells which are larger than 6 inches in diameter

TTV = 1.5 x SWV

Table C.5. Chlorine Dosage Formula

METHOD 1

Standing Well Volume (SWV)

SWV = gal/ft x feet of water

Total Disinfectant Volume (TDV)

TDV = 4 x SWV

$$\text{Hypochlorite Gallons} = \frac{\text{TDV x \% Chlorine Desired}}{\text{\% Active Hypochlorite Solution}}$$

Example: Well Diameter = 6 inches
Static Water Level = 19 feet
Well Depth = 140 feet
Chlorine Level Desired = 200 mg/l
Sodium Hypochlorite Activity = 10%

Chlorine Value	% Value
50 mg/l	0.005%
200 mg/l	0.02%
500 mg/l	0.05%
1,000 mg/l	0.1%

SWV = 1.47 gal/ft x 121 feet = 178 gallons
TDV = 4 x 178 gals = 712 gallons

$$\text{Hypochlorite Gals} = \frac{\text{712 gals x .02\%}}{\text{10\%}} = \text{1.42 gals of 10\% Sodium Hypochlorite}$$

NOTE: The TDV is the total volume of chlorine solution required for disinfection and is sufficient to flood the well and surrounding zone.

METHOD 2

(more exact, based on weight)

10% Hypochlorite = 9.6 lbs/gal
Weight of Water = 8.34 lbs/gal

$$\frac{\text{Wt of TDV} \times \text{Desired Chlorine}}{\text{\% Al Hypochlorite} \times \text{Wt of Hypochlorite}} = \text{Gallons of Hypochlorite}$$

TDV = 712 gals x 8.34 = 5,938 pounds

$$\frac{5{,}938 \text{ lbs} \times .02\%}{10\% \times 9.6 \text{ lbs/gal}} = 1.24 \text{ gallons of 10\% Hypochlorite}$$

Table C.6. Formulae for calculating pH control for chlorination

Using NuWell-310 or QC-21

$$\text{Gallons NuWell-310 or QC-21} = \frac{\text{Alk}}{100\text{ mg/l}} \times \frac{\text{Chlorine}}{200\text{ mg/l}} \times \frac{\text{Volume}}{2{,}000\text{ gals}}$$

Example: Water alkalinity (Total) = 150 mg/l
Chlorine level desired = 200 mg/l
SWV = 900 gals TDV = 3,600 gals

$$\frac{150}{100} \times \frac{200}{200} \times \frac{3{,}600}{2{,}000} = 2.7 \text{ gallons of NuWell-310 or QC-21}$$

Vinegar: 1 gal of vinegar will adjust 100 gals of water where average alkalinity is 100 mg/l and the chlorine dose is 500 mg/l

$$\text{Gallons of Vinegar} = \frac{\text{Alkalinity}}{100\text{ mg/l}} \times \frac{\text{Desired Chlorine levels}}{500\text{ mg/l}} \times \frac{\text{Volume}}{100\text{ gals}}$$

Hydroxyacetic: (Glycolic) 1qt (liter) will adjust 1,000 gals of water with an average alkalinity of 100 mg/l and a chlorine dose of 200 mg/l

$$\text{Quarts of Hydroxyacetic} = \frac{\text{Alkalinity}}{100\text{ mg/l}} \times \frac{\text{Desired Chlorine levels}}{200\text{ mg/l}} \times \frac{\text{Volume}}{1{,}000\text{ gals}}$$

Table C.7. Oxidation state of elements

Element	Symbol	Atom No.	Atomic Wt.	Oxidation State
Aluminum	Al	13	26.981	+3
Antimony	Sb	51	121.75	±3, +5
Arsenic	As	33	74.921	±3, +5
Barium	Ba	56	137.327	+2
Beryllium	Be	4	9.012	+2
Bismuth	Bi	83	208.980	+3, +5
Boron	B	5	10.811	+3
Bromine	Br	35	79.904	±1, +5
Cadmium	Cd	48	112.411	+2
Calcium	Ca	20	40.078	+2
Carbon	C	6	12.011	+2, ±4
Cesium	Cs	55	132.905	+1
Chlorine	Cl	17	35.452	±1, +5, +7
Chromium	Cr	24	51.996	+2, +3, +6
Cobalt	Co	27	58.933	+2, +3
Copper	Cu	29	63.546	+1, +2
Florine	F	9	18.998	-1
Gallium	Ga	31	67.723	+3
Gold	Au	79	196.966	+1, +3
Helium	He	2	4.002	0
Hydrogen	H	1	1.007	±1
Iodine	I	53	126.904	±1, +5, +7
Iron	Fe	26	55.847	+2, +3
Lead	Pb	82	207.2	+2, +3
Lithium	Li	3	6.941	+1
Magnesium	Mg	12	24.305	+2
Manganese	Mn	25	54.938	+2, +3, +4, +7
Mercury	Hg	80	200.59	+1, +2
Molybdenum	Mo	42	95.4	+6

Element	Symbol	Atom No.	Atomic Wt.	Oxidation State
Nickel	Ni	28	58.69	+2, +3
Nitrogen	N	7	14.006	±1, 2, 3, +4, +5
Oxygen	O	8	15.999	-2
Palladium	Pd	46	106.42	+2, +4
Phosphorus	P	15	30.973	±3, +5
Platinum	Pt	78	195.08	+2, +4
Plutonium	Pu	94	244.06	+3, 4, 5, 6
Potassium	K	19	39.098	+1
Radium	Ra	88	226.025	+2
Radon	Rn	86	222.017	0
Selenium	Se	34	78.96	+4, +6, -2
Silicon	Si	14	28.085	+2, ±4
Silver	Ag	47	107.868	+1
Sodium	Na	11	22.989	+1
Strontium	Sr	38	87.62	+2
Sulfur	S	16	32.066	+4, +6, -2
Tellurium	Te	52	127.60	+4, +6, -2
Thallium	Tl	81	204.383	+1, +3
Thorium	Th	90	232.038	+4
Tin	Sn	50	118.710	+2, +4
Titanium	Ti	22	47.88	+2, +3, +4
Tungsten	W	74	183.85	+6
Uranium	U	92	238.028	+3, 4, 5, 6
Vanadium	V	23	50.941	+2, 3, 4, 5
Xenon	Xe	54	131.29	0
Zinc	Zn	30	65.39	+2
Zirconium	Zr	40	91.224	+4

References

Amy PS, Haldeman DL, editors. 1997. The microbiology of the terrestrial deep subsurface. Boca Raton: Lewis Publishers. 356 p.

Applegate LE, Erkenbrecher CW. 1989. New chloramines process to control aftergrowth and biofouling in Permasep B-10 RO surface seawater plants. Desalination 74:51–67.

Baird JO. 2002. Personal communication.

Brewster RQ, McEwen WE. 1963. Organic chemistry. 3rd ed. Englewood Cliff: Prentice Hall Inc. 854 p.

Chapelle FH. 1993. Groundwater microbiology and geochemistry. New York: John Wiley and Sons. 424 p.

Characklis WG, Marshall KC, editors. 1990. Biofilms. New York: John Wiley and Sons. 584 p.

Characklis WG. 1980. Biofilm development and destruction. Houston: Rice Univ Pr. 246 p.

Clark JA, Pagel JE. 1977. Pollution indicator bacteria associated with municipal raw and drinking water supplies. Can Jour Micro 23:465–78 p.

Cullimore DR. 1992. Practical manual of groundwater microbiology. Chelsea, MI: Lewis Publishers. 412 p.

Driscoll FG, editor. 1986. Groundwater and wells. St. Paul: Johnson Screens. 1089 p.

Fenchel T, King GM, Blackburn TH. 1988. Bacterial biogeochemistry, the ecophysiology of mineral cycling. London: Academic Pr. 307 p.

Flemming HC, Wingender JW. 2001. Structural ecological and functional aspects of EPS. In: Gilbert P, Allison D, Brading M, Verran J, Walker J, editors. Biofilm community interaction, chance or necessity. Cardiff, UK: Bioline. p 175–89.

Fowkes FM. 1965. Attractive forces and interfaces. In: Chemistry and Physics of Interfaces. Washington: ACS Pr.

Hem JD. 1985. Study and interpretation of the chemical characteristics of natural water. USGS Water Supply Paper 2254. Washington: US Gov Print Office. 263 p.

Holben RG. 2001. Personal communication.

Holben RJ, Gaber M. 2002. On site water supply disinfection manual. Lansing: Michigan Department of Environmental Quality: 55 p.

Holt JG, Krieg NR, Sneath PH, Staley JT, Williams ST, editors. 1994. Bergey's manual of determinative bacteriology. 9th ed. Baltimore: Williams & Wilkins. 787 p.

Hurst CJ, Crawford RL, Knudson GR, McInerney MJ, Stetzebbach LD, editors. 2002. Manual of environmental microbiology. 2nd ed. Washington: ASM Pr. 1138 p.

Jass J, Wai SN. 2000. Vibrio cholorae biofilm at the air and solid, liquid interface. Big Sky (MT): ASM Conference Presentation.

Karstgens V, Flemming HC, Wingender J, Borchard W. 2000. Influence of calcium ion concentration on the mechanical properties of a model biofilm of *Pseudomonas aeruginosa.* Water Science Technology 43(1):49–57.

Klahre J, Flemming HC. 2000. Monitoring of biofilm in paper mill water system. Water Research 34:3657–65.

Langelier WF. 1936. The analytical control of anticorrosion water treatment. AWWA Journal 28:1500–21.

Lappin-Scott HM, Bass CJ, McAlpine K, Sanders PE. 1994. Survival mechanisms of hydrogen sulfide producing bacteria isolated from extreme environments and their role in corrosion. Internat'l Biodeg and Biodeter 34:305–19.

LeChevallier MW, Bobcock TM. 1987. Examination and characterization of distribution system biofilms. Appl Env Micro 53(12): 2714–2724.

Lehr J, Hurlburt S, Gallagher B, Voytek. 1988. Design and construction of water wells. New York: Van Nostrand Reinhold. 229 p.

Manahan SE. 1994. Environmental chemistry. 2nd ed. Boca Raton: Lewis Publishers. 811 p.

Mansuy N. 1999. Water well rehabilitation, a practical guide to understanding well problems. Boca Raton: Lewis Publishers. 174 p.

Marsh PD, Bradshaw DJ. 1996.Organisms and multispecie biofilms. In: Proc Microbial Biofilm ASM Conference. Washington: ASM Pr. 42 p.

Mehmert M. 1994. Why agitation improves chemical cleaning. National Driller Buying Guide (12):27.

Mehmert M. 1995. How bacteria complicate well rehabilitation. WWJ 7:65–68.

Molin S. 2000. Direct in-situ observation of the metabolic activity of bacteria growing in biofilm. In: Proc Biofilms 2000 ASM Conference: Washington. ASM Pr. 98 p.

Neu T. 1996. Significance of bacterial surface active compounds in interaction of bacteria with interfaces. Microbiological Review 60:151–66.

Pontius FW. Editor. 1990. Water quality and treatment. 4th ed. New York: AWWA-McGraw Hill Inc. 1194 p.

Rose JB, Grimes DJ. 2000. Reevaluation of microbial water quality: powerful new tools for detection and risk assessment. Florida: Amer Acad Micro. 21 p.

Schnieders JH. 2001. Oct. Coliforms and disinfection of water wells. WWJ 55(15):2–15.

Schnieders JH. 1996. Dispersant chemistry in water well cleaning. In Proc AWWA Water Technology Conference. Denver: AWWA Pr.

Schnieders JH. 1998 Feb. Well Chlorination. WWJ 52(2):25–26.

Schnieders MJ, Adkinson C. 2000. Taking a new look at an old process: chemical cleaning of a filtration water plant. Opflow, JAWWA 26(10):12–15.

Schnieders MJ. 2001. When the well runs dry: keeping fouling in check.Water Conditioning and Purification. 43(12):32–34.

Sienko MJ, Plane RA. 1957. Chemistry. New York: McGraw Hill Book Co. 621 p.

Smith SA. 1995. Monitoring and remediation wells,problem preventing, and remediation. Boca Raton: Lewis Publishers. 183 p.

Spath R, Flemming HC, Wuertz S. 1998. Sorption properties of biofilm. Water Science Technology 37:207–210.

Sutherland IW. 2001. Biofilm exopolysaccharides: a strong and sticky framework. Microbiology 147(1):3–9.

Troutt JR. 2002. Personal communication.

Van der Kooij D, Hijnen WAM. 1985. Measuring the concentration of easily assimilable organic carbon (AOC) treatment as a tool for limiting regrowth of bacteria in distribution systems. Proc AWWA Water Quality Tech Conf. Houston: AWWA Pr.

Van der Kooij D, Hijnen WA, Kruithof JC. 1989. The effects of ozonation, biological filtration, and distribution on the concentration of easily assimilable organic carbon (AOC) in drinking water. Ozone Science Engineer 11:297–311.

Whistler RL, Schweiger K. 1957. Oxidation of amylopectin with hypochlorite at different hydrogen ions concentration. JACS 79(24)6460–64.

Glossary

***Acinetobacter*.** Aerobic, slime forming bacteria, which use organic carbon as a food source (heterotrophic). They are normally found in the soil and when inhabiting a system with *Pseudomonas* tend to cause increased biofilm growth.

***Actinomycetes* Group.** Aerobic bacteria that form branching filaments. Usually only found in idle wells. May inhabit the borehole wall and cause severe blockage.

aerobic. An organism that grows in the presence of oxygen.

agglomeration. To cluster together to form in a rounded mass.

amphoteric. A characteristic of a substance that it is active both in the acid and alkaline pH range.

anaerobes. A bacteria that lives without oxygen, usually has a fermentative metabolism (breaks sugars and other complex organic material into simple organic acids and hydrogen) and contains most members of the sulfate-reducing bacteria (SRB).

anaerobic. Organisms that can grow in the absence of oxygen or an environment without oxygen.

anionic. Negatively charged.

anoxic. Without oxygen.

biodispersant. A dispersant that is active against biopolymers. Several of the commercial products such as NW-310, QC-21, and UNICID Catalyst also have dispersant activity against minerals.

bioenvironment. An environment conducive to microorganism growth.

biofilm. A film of microbial growth which consist of the bacterial cells and the slimy exopolymer which binds them to the surface. The habitat of bacteria.

biopolymers. Polymeric product like produced by bacteria or other organisms.

brucite. Magnesium hydroxide.

by-product. A product produced while producing something else, usually a chemical substance produced as a result of a standard chemical or biological activity.

calcite. Calcium carbonate.

calcium hypochlorite. A crystal or powder form of hypochlorite having the calcium or hardness ion as part of the compound.

cationic. Positively charged.

cfu (colony forming units). When spread on a plate of agar, each bacterium or group of bacteria will form a visible colony. These colonies are counted and the count is used to evaluate the number of bacteria originally present in the water sample.

chelate. A compound that is soluble in water and combines with metal ions to keep them in solution.

coaggregation. The transfer of genetic information from one bacteria species to another, as an example the ability to degrade a certain chemical.

coliform. An aerobic slime former characterized by their fermentation of lactose and used as an indicator of waste water contamination. It can also live anaerobically and can be found in the intestinal tract of animals, but is also found in nature.

colloid. A suspended solid usually less than one micron in diameter which can not be filtered easily and will not settle out of solution rapidly.

contaminant. Any chemical or microbial agent foreign to a well. Typical mineralization or biological accumulation in wells are not generally referred to as contaminants.

decontamination. The cleaning process used to remove harmful substances such as infectious agents or harmful chemicals to reduce the possibility of disease or harm from those substances.

disinfection. The selective destruction of disease causing microorganisms.

dispersant. An additive that prevents agglomeration of particulates, or is used to break up concentrations of mineral ions or organic matter such as biofilm.

dolomite. A carbonate formation consisting of calcium and magnesium carbonates, usually in a 50:50 ratio.

E. coli (Escherichia coli). The most famous member of the fecal coliform group. It comprises the dominant bacteria in the human feces.

environmentally active organism. An organism that thrives in a natural environment.

exopolymer or extracellular polymeric substance (EPS). These are polymeric material, usually polysaccharides, which are synthesized within the bacterial cell and secreted to form a capsule or slimy matrix around the cell to aid in attachment or to protect the cell or as a nutrient capture system. It forms the building blocks for biofilm.

exothermic. A substance or reaction that releases heat.

extracellular polymeric substance (EPS). Polysaccharide material produced by bacteria (slime).

facultative. The ability for an organism to live with or without oxygen.

fastidious. Having complicated nutritional requirements.

ferric oxyhydroxide. A compound of oxidized iron which is formed following the oxidation of iron by iron oxidizing bacteria. The material usually has a slick or slimy feel and is red in color.

Flavobacter. These aerobic slime formers are able to utilize many kinds of organic compounds and are widely distributed in nature. Because of their wide distribution they are often found in aerobic aquifers (between 3 and 10%), however they have not been reported isolated from cored aquifer sediment.

gravel pack. Also called filter pack. It is sand or gravel that is smooth, uniformed, cleaned, well rounded, and siliceous. It is placed in the annulus of the well between the borehole wall and the well screen to prevent formation material from entering the screen.

gypsum. Calcium sulfate.

heterotrophic. Organisms that are able to utilize organic material as the principal source of energy and carbon for growth.

homeothermic (homoiothermic). Maintaining a constant warm body temperature – warm blooded.

host organisms. Any organism on or in, which another organism lives for nourishment, development, or protection.

HPCs. Heterotrophic plate count usually reported in colony forming units (cfu) per milliliter. Often used to assess the bacterial load in a water sample.

hydraulic conductivity. The rate of flow of water in gallons per day through a cross section of one sq. ft. under a unit hydraulic gradient, at the prevailing temperature (gpd/ft^2).

hydrophilic. Having an affinity for, absorbing, wetting smoothly with, tending to combine with or capable of dissolving in water.

hydrophobic. Tending not to combine with or incapable of dissolving in water.

inoculum. A volume used to implant microorganisms into a culture medium.

ions. An ion is an atom which has gained or lost an electron, usually a product is dissolved such as the dissolving of salt produces two separate ions: sodium, a positive ion, and chlorine, a negative ion.

iron oxidizing bacteria. Any bacteria which can oxidize ferrous iron.

jetting. The act of projecting a stream of water or gas, or a mixture at a surface for cleaning effect.

Langelier Saturation Index. A calculated saturation index for calcium carbonate, useful in predicting the deposit forming character of water.

microstructures. In wells the small structures formed by the action of mineral deposition and bacterial growth in the well and the formation.

mineralization. The depositing of insoluble minerals such as calcium carbonate and calcium sulfate onto well surfaces or in pore spaces in the well environment.

miscible. Capable of being mixed.

mitosis. The process by which bacteria reproduce by dividing into two daughter cells.

non-enteric. Referring to an organism not from the intestines.

NTU. Nephelometric turbidity unit. Used to measure the turbidity of water.

oxides. A compound or substance formed when a metal ion bonds with oxygen, such as iron with oxygen producing iron oxide.

oxidizer. A species that takes electrons from a reducing agent in a chemical reaction.

passivation. Formation of a thin coating on a metal surface, usually an oxide film which prevents further oxidation or corrosion.

pathogen. A microbe that causes disease.

piezometric head. The difference measured between two separate readings of the water table elevation as observed in a piezometer (a small diameter pipe open at the bottom inserted into an aquifer to observe water table changes).

planktonic. A microorganism that is free swimming or moving freely in the water.

polymer. The chemical formed when small molecules called monomers bond together in a repeatable configuration to form a much larger molecule. Many natural substances form polymers, such as the cellulose in woods and papers or sugars. Many synthetics polymers are the basis of our textile, plastic, rubber, and detergent industries.

polysaccharide polymer. The primary chemistry of the exopolymer produced by bacteria as a slimy coating or main constituent of biofilm. It is a polymer made up of sugars.

pore spaces. The space or openings between formation structure that allow water flow.

precipitates. The solids which drop out of solution because their constituents reach a saturation point or excess concentration.

protozoans. Single celled microscopic organism considered the most primitive form of animal life. Organisms like the amoebas, flagellates, and ciliates make up this group.

Pseudomonas. Probably the most universal slime former. It is an aerobic bacteria which uses a wide selection of organic materials for its energy and carbon source. Many of the different species are facultative often using alternative path ways for respiration when oxygen concentrations becomes too low. Some species can be pathogenic.

reductive. To receive an electron, to become more negative.

reducing agent. A species that supplies electrons during a chemical reaction.

rehabilitation. The complete cleaning of a well including reworking the pump and chemical cleaning the casing, screen, and gravel pack.

saponify. Hydrolysis or conversion of fats or oils by an alkali to a more water soluble form.

sequestering. Inhibiting normal ion behavior by combining one of the active ions with added material.

sessile. Attached to a surface or incorporated into a biofilm so that the organisms are not washed into the flow.

sodium hypochlorite. The sodium form of hypochlorite which supplies hypochlorite above 6.5 pH and hypochlorous acid below 6.5 pH.

stoichiometric. Methodology by which quantities of reactants and products in chemical reactions are determined.

sulfur oxidizing bacteria. Aerobic bacteria that oxidize sulfides to sulfate.

sulfate-reducing bacteria (SRB). Anaerobic bacteria which reduces sulfate and produces hydrogen sulfide gas.

surfactant. A surface active agent which reduces the friction of water moving over surfaces.

surging. The act of moving the well water in and out of the acquifer by moving a surge block up and down in the casing or screen area.

symbiotic. Relationship between two organisms that does not harm either and may be beneficial to one or both.

Thiobacillus. Sulfur oxidizing bacteria that are known for sulfuric acid production and low pH.

Thiothrix. Sulfur oxidizing bacteria responsible for producing sulfate accumulation and known by their characteristic rosetta formation. They oxidize sulfides.

toxic. Harmful, destructive or deadly.

tremied. Act of injecting chemicals into a well through a small diameter line.

trihalomethanes (THMs). A type of disinfectant by-product formed when chlorine reacts with organic substances in water.

Index

C

D

E

M

N

O

P

Q

R

S

T

U

V

W

Y